BIOLOGICAL AND MEDICAL PHYSICS, BIOMEDICAL ENGINEERING

BIOLOGICAL AND MEDICAL PHYSICS, BIOMEDICAL ENGINEERING

The fields of biological and medical physics and biomedical engineering are broad, multidisciplinary and dynamic. They lie at the crossroads of frontier research in physics, biology, chemistry, and medicine. The Biological and Medical Physics, Biomedical Engineering Series is intended to be comprehensive, covering a broad range of topics important to the study of the physical, chemical and biological sciences. Its goal is to provide scientists and engineers with textbooks, monographs, and reference works to address the growing need for information.

Books in the series emphasize established and emergent areas of science including molecular, membrane, and mathematical biophysics; photosynthetic energy harvesting and conversion; information processing; physical principles of genetics; sensory communications; automata networks, neural networks, and cellular automata. Equally important will be coverage of applied aspects of biological and medical physics and biomedical engineering such as molecular electronic components and devices, biosensors, medicine, imaging, physical principles of renewable energy production, advanced prostheses, and environmental control and engineering.

D. Shi (Ed.)

Biomaterials and Tissue Engineering

With 72 Figures and 22 Tables

Springer

Professor Donglu Shi
University of Cincinnati
Department of Chemical
and Materials Engineering
493 Rhodes Hall
Cincinnati, OH 45221-0012
USA
e-mail: shid@email.uc.edu

ISSN 1618-7210

ISBN 3-540-22203-0 Springer Berlin Heidelberg New York

Library of Congress Control Number: 2004106906

Springer is a part of Springer Science+Business Media.

springeronline.com

Printed in Germany

Cover concept by eStudio Calamar Steinen
Cover production: *design & production* GmbH, Heidelberg

Printed on acid-free paper SPIN 10844595 57/3141/di - 5 4 3 2 1 0

Preface

The current interest in developing novel materials has motivated an increasing need for biological and medical studies in a variety of clinical applications. Indeed, it is clear that to achieve the requisite mechanical, chemical and biomedical properties, especially for new bioactive materials, it is necessary to develop novel synthesis routes. The tremendous success of materials science in developing new biomaterials and fostering technological innovation arises from its focus on interdisciplinary research and collaboration between materials and medical sciences. Materials scientists seek to relate one natural phenomenon to the basic structures of the materials and to recognize the causes and effects of the phenomena. In this way, they have developed explanations for the changing of the properties, the reactions of the materials to the environment, the interface behaviors between the artificial materials and human tissue, the time effects on the materials, and many other natural occurrences. By the same means, medical scientists have also studied the biological and medical effects of these materials, and generated the knowledge needed to produce useful medical devices.

The concept of biomaterials is one of the most important ideas ever generated by the application of materials science to the medical field. In traditional materials research, interest focuses primarily on the synthesis, structure, and mechanical properties of materials commonly used for structural purposes in industry, for instance in mechanical parts of machinery. The evolution of the field has rapidly crossed many boundaries and interfaced with other interdisciplinary research, among which biomedical engineering, biochemistry, and the medical sciences are the most active fields, especially in the 21st century. In the development of biomaterials, researchers must not only deal with the basic problems in synthesis and characterization; new challenges are also presented by the biological behaviors of these novel materials. In particular, what kind of interface structures will form when an artificial material comes to contact with a human soft or hard tissue? Is the artificial material introducing harmful effects to a human body? Is it possible to develop more biologically compatible materials that are suitable for certain medical devices used in the human body? How are we going to quantifiably define biocompability of a material? Can a material be synthesized to be "bioactive?" Are there fundamental relationships between materials lattice structures and bioactivity? These are some of the new questions that must be answered jointly by materials scientists and medical researchers. Indeed, the unique characteristics of biomaterials research mean that it requires extensive collaboration between researchers in both fields.

A more fundamental issue concerns teaching in these disciplinary fields. Today, the teaching of biomaterials in our schools remains conventional. The current curricula still concentrate largely on structural materials. University faculty members tend to teach their courses based on the traditional concepts of materials. One reason for this is the lack of suitable biomaterials textbooks accessible to both undergraduate and graduate students. Most of the existing books adopt a traditional manner of introducing materials and/or fail to address the critical issues related to biological phenomena. While biomaterials research is actively pursued collaboratively by both materials and medical scientists, the teaching lags far behind partly due to this lack of comprehensive and systematic textbooks and monographs in the field.

Biomaterials and Tissue Engineering and its companion volume *Medical Devices and Applications* should help to fill this gap in the literature. This book is addressed to the teachers and researchers in a broad spectrum of materials, chemical, biological and medical sciences and engineering. It summarizes the wealth of experimental results in both biomaterials and tissue engineering and introduces new aspects of materials science relevant to biological and medical applications. It also demonstrates new trends in the field and presents novel methods and techniques for experiments and for the development of unique biomaterials. We are grateful to all invited authors for their excellent contributions to this book.

Donglu Shi
Professor
University of Cincinnati
June, 2003

Contents

List of Contributors

Donglu Shi

Department of Materials Science and Engineering, University of Cincinnati
Cincinnati, OH 45221-0012, USA
Email: Dshi@ uceng. uc. edu

Alexis Clare

Director, Industry-University Center for Bioceramics
Alfred University, USA
Email: clare@ alfred. edu

Min Wang

Medical Engineering and
Mechanical Engineering
Faculty of Engineering
University of Hong Kong
Pokfulam Road, Hong Kong
Email: minwangbiomats@ yahoo. co. uk

Toshihiro Kasuga

Graduate School of Materials
Science and Engineering
Nagoya Institute of Technology
GoKiso-Cho, Showa-Ku
Nagoya 466-8555, Japan
Email: kasugato@ mse. nitech. ac. jp

Miqin Zhang

302L Roberts Hall
Department of Materials Science
and Engineering University of
Washington
Seattle, WA 98195-2120, USA
Email: mzhang@ u. washington. edu

Qing Liu

Celgene Cellular Therapeutics
Division 45 Horse Hill Road,
Suit 110 Cedar Knolls,
NJ 07927 USA
Email: qingliu@ yahoo. com

1 Bioactive Materials and Processing

Min Wang

1.1 Introduction

It is widely recognized that the rapid and continuing change in emphasis in materials science away from traditional engineering materials has been largely instituted by the requirements of emerging technologies for advanced and structurally sophisticated new materials. Medical engineering, often included in the list of advanced technologies, requires the underpinning of high-tech materials. The word which is used to categorize materials for biomedical applications is "biomaterial".

A biomaterial is a synthetic material used to replace part of a living system or to function in intimate contact with living tissue (Ratner et al., 1996). The Clemson University Advisory Board for Biomaterials defined a biomaterial as "a systemically and pharmacologically inert substance designed for implantation within or incorporation with living systems". In 1986, the Consensus Conference of the European Society for Biomaterials defined a biomaterial as "a nonviable material used in a medical device, intended to interact with biological systems" (Williams, 1987a). Another definition of biomaterial is "any substance (other than drugs) or combination of substances synthetic or natural in origin, which can be used for any period of time, as a whole or as a part of a system which treats, augments, or replaces any tissue, organ, or function of the body" (Hulbert et al., 1987; Helmus and Tweden, 1995). A biomaterial is different from a biological material such as bone which is produced by a biological system.

The really significant part of the Consensus Conference definition is that it has referred to the interaction with biological systems. To qualify for this descriptive definition, a material must be used in a situation in which it is to remain in contact with living systems for a sufficiently long time for some significant interaction to take place. These situations largely involve implantation in the body but can also be extracorporeal circuits (e.g., kidney machines) and devices having intentionally prolonged contact with external surfaces (e.g., contact lenses and wound dressings). To date, biomaterials are used clinically or experimentally in implantable electronic devices, drug delivery systems, hybrid artificial organs, bone substitutes, ligament and tendon replacements, extracorporeal blood separation columns, and so on.

The UK Office of Science and Technology (OST) in 1995 identified biomaterials as one of the eight priority areas within the sector of materials: i. e. a new generation of materials which encourage and enhance the restoration and repair of body tissue function (UK OST Report, 1995). "Biomaterials save lives, relieve suffering and improve the quality of life for a large number of patients every year" (IoM Report, 1995). In order to develop and use materials in medicine, one must have sufficient knowledge of different disciplines and collaborate with people of various specialties. Knowledge in the following three areas is essential:

(1) Materials science and engineering: processing – structure – property interrelationship of synthetic and biological materials, including metals, ceramics, polymers, composites, tissues, etc..

(2) Biology and physiology: cell and molecular biology, anatomy, animal and human physiology, etc..

(3) Clinical sciences: dentistry, ophthalmology, orthopedics, plastic and reconstructive surgery, cardiovascular surgery, neurosurgery, immunology, histopathology, experimental surgery, veterinary medicine and surgery, etc..

The role of biomaterials has been influenced considerably by advances in many areas of science and technology. Biomaterials can be classified in a number of ways according to different criteria (Black, 1992; Greco, 1994; Park and Lakes, 1992). The classification of biomaterials as polymers, metals, ceramics, and composites, as shown in Table 1. 1, is generally adopted.

Table 1. 1 Materials for use in the body

Material	Advantage	Disadvantage	Application
Polymers			
Nylon	Ductile,	Not strong,	Suture,
PTFE	light,	prone to creep,	vascular prosthesis,
Polyester	easy to fabricate	degradable	accetabular cup,
Silicone			artificial ligament
Metals			
Ti and its alloys	Ductile,	Prone to corrosion,	Artificial joint,
Co – Cr alloys	strong,	unwanted ion release	bone plate and screw,
Stainless steels	tough		dental root implant,
Au, Ag, Pt			pacer,
			suture wire
Ceramics			
Carbon	Biocompatible,	Brittle,	Cardiovascular device,
Aluminum oxide	Inert or bioactive,	weak in tension,	dental prosthesis,
Hydroxyapatite	strong in compression,	sometimes fragile	joint prosthesis,
	stiff		orthopedic implant
Composites			
Carbon-carbon	Strong,	Difficult to make,	Joint implant,
Metal – PMMA	stiff,	high production cost	heart valve,
HA – HDPE	tailor-made,		bone cement
	distinctive properties		

According to Hench and Ethridge, "a bioactive material is one that elicits a specific biological response at the interface of the material which results in the formation of a bond between the tissues and the material" (Hench and Ethridge, 1982). At present, bioactive materials include some calcium phosphate compounds, bioactive glasses, bioactive glass-ceramics, bioactive ceramic coatings deposited on metal substrates, composites containing bioactive ceramic phase(s), etc. A characteristic feature common to these materials is that they bond to human bone with no fibrous tissue at the interface.

1.2 Calcium Phosphate Ceramics

Calcium phosphate bioceramics have been in use in medicine and dentistry for more than 20 years. The interest in one group member, hydroxyapatite, arises from its similarity to bone apatite, the major component of the inorganic phase of bone, which plays a key role in the calcification and resorption processes of bone (Fawcett, 1986). In the mid-1970s, three groups, Jarcho et al. in the USA, de Groot et al. in Europe, and Aoki et al. in Japan, simultaneously but independently worked toward the development and commercialiation of hydroxyapatite as a biomaterial for bone repair, augmentation, and substitution.

Different phases of calcium phosphate ceramics can be used in medicine, depending on whether a bioactive or a resorbable material is desired. Table 1.2 lists calcium phosphates that are often encountered in research and clinical in use.

Table 1.2 Family of calcium phosphate compounds

Mineral Name	Chemical Name	Chemical Formula	Ca : P (Molar Ratio)
Monetite	Dicalcium phosphate (DCP)	$CaHPO_4$	1.00
Brushite	Dicalcium phosphate dihydrate (DCPD)	$CaHPO_4 \cdot 2H_2O$	1.00
	Octocalcium phosphate (OCP)	$Ca_8(HPO_4)_2(PO_4)_4 \cdot 5H_2O$	1.33
Whitlockite		$Ca_{10}(HPO_4)(PO_4)_6$	1.43
	Tricalcium phosphate (TCP)	$Ca_3(PO_4)_2$	1.50
Hydroxyapatite	Hydroxyapatite (HA)	$Ca_{10}(PO_4)_6(OH)_2$	1.67
Hillinstockite	Tetracalcium phosphate (TTCP)	$Ca_4P_2O_9$	2.00

The stable phases of calcium phosphate ceramics depend considerably on the temperature and the presence of water, either during materials processing or in the in-service environment. At body temperature, only two calcium phosphates are stable when in contact with aqueous media such as body fluids. At $pH < 4.2$, the stable phase is $CaHPO_4 \cdot 2H_2O$ (brushite), while at $pH > 4.2$, the stable phase is $Ca_{10}(PO_4)_6(OH)_2$ (hydroxyapatite). At higher temperatures, other phases such as $Ca_3(PO_4)_2$ (TCP) and $Ca_4P_2O_9$ (TTCP) are present. The unhydrated high-temperature calcium phosphate phases interact with water or body fluids at 37℃ to form hydroxyapatite (HA).

1.2.1 Biological Apatites

Biological apatites constitute the mineral phase of calcified tissues such as bone, dentine, and enamel in the body and also some pathological calcifications. They are similar to synthetic HA, but they differ from HA in composition, stoichiometry, and physical and mechanical properties. Biological apatites are usually calcium-deficient as a result of various substitutions at regular HA lattice points. It is therefore not appropriate to simply refer to biological apatite as hydroxyapatite.

1.2.1.1 Composition and Structure

The general chemical formula for biological apatites is

$$(Ca, M)_{10}(PO_4, CO_3, Y)_6(OH, F, Cl)_2 \tag{1.1}$$

where M represents metallic elements such as Na, K, and Mg; and Y represents functional groups such as acid phosphate, sulfates, etc. As compared to synthetic HA with the chemical formula of $Ca_{10}(PO_4)_6(OH)_2$, substitution in biological apatites by the carbonate group for the phosphate group takes place in a coupled manner, i.e. CO_3 for PO_4 and Na for Ca. The coupled substitution is necessary to balance charges for the substitution. Differences in composition among apatites in human enamel and bone and HA are shown in Table 1.3.

Table 1.3 Composition of biological apatites and hydroxyapatite

Major Constituent	Biological Apatite		Hydroxyapatite (wt%)
	In enamel (wt%)	In bone (wt%)	
Ca	36.00	24.50	39.60
P	17.70	11.50	18.50
Na	0.50	0.70	
K	0.08	0.03	
Mg	0.44	0.55	
F	0.01	0.02	
Cl	0.30	0.10	
CO_3^{2-}	3.20	5.80	
Trace elements: Sr, Pb, Ba, Fe, Zn, Cu, etc.			
Ca:P (molar ratio)	1.62	1.65	1.67

1.2.1.2 Property

Because of substitutions and hence differences in composition, biological apatites possess different physical and mechanical properties, compared to synthetic HA. Table 1.4 lists the properties of apatites in human enamel and bone and HA.

Table 1.4 Properties of biological apatites and hydroxyapatite

Property	Biological Apatite		Hydroxyapatite
	In enamel	In bone	
Lattice parameter /nm			
a	0.9441	0.9419	0.9422
c	0.6882	0.6880	0.6880
Crystal size /nm	130 × 30	25 × (2.5 – 5.0)	in micrometers
Elastic modulus /GPa	14	7 – 30	10
Tensile strength /MPa	70	50 – 150	~ 100

The biological apatite of enamel differs from those of dentine and bone in crystallinity and in the concentration of minor elements. It has the largest apatite crystal size compared to dentine and bone apatites and it is the hardest tissue in the human body. Enamel apatite is less soluble than dentine or bone apatite but is much more soluble than dense HA. X-ray diffraction (XRD) patterns and infrared (IR) spectra of biological apatites (in bone, enamel, and dentine) are distinctively different from those obtained from synthetic HA.

Sintering products of biological apatites of enamel, dentine, and bone are different, which indicates the differences in their composition and calcium deficiency (LeGeros, 1990). Sintering human enamel and dentine apatites above 800℃ yielded hydroxyapatite and small amounts of β-tricalcium phosphate (β-TCP). Sintering human bone apatite above 800℃ gave hydroxyapatite (HA) and minor amounts of CaO. Sintering animal bones resulted in the formation of β-TCP or CaO depending on the age and species of animal (LeGeros, 1991).

1.2.2 Hydroxyapatite

Hydroxyapatite belongs to the apatite family. Apatite is the name given to a group of crystals of the general chemical formula $M_{10}(RO_4)X_2$, where R is most commonly phosphorus, M could be one of several metals, although it is usually calcium, and X is commonly hydroxide or a halogen such as fluorine or chlorine. Possible M, R and X are listed below (Aoki, 1994):

$$M = Ca, Sr, Ba, Cd, Pb, Mg, Na, K, etc.$$
$$R = P, CO_3, V, As, S, Si, Ge, Cr, B, etc.$$
$$X = OH, CO_3, O, BO_2, F, Cl, Br, vacancy, etc.$$

The mineral, first called apatite in 1788 by Werner, had been confused earlier with other minerals such as aquamarine, olivine, and fluorite (Deer et al., 1985). Apatite minerals are found in almost all igneous rocks as well as in sedimentary and metamorphic rocks.

1.2.2.1 Structure

Hydroxyapatite is the most commonly used calcium phosphate in the medical field, as it possesses excellent biocompatibility and is osteoconductive. It has the chemical formula $Ca_{10}(PO_4)_6(OH)_2$ and a Ca/P molar ratio of 1.67. Its theoretical density is 3.156 g/cm^3. HA crystal has hexagonal rhombic prisms, with the space group being $P6_3/m$. This space group is characterized by a sixfold c-axis perpendicular to three equivalent a-axes (a_1, a_2, and a_3) at angles of 120° to each other. Its unit cell contains a complete representation of the apatite crystal, consisting of Ca^{2+}, PO_4^{3-}, and OH^- groups. The 3-D crystal structure of HA is shown in Fig. 1.1 (Aoki, 1994). The structure of HA projected on the c axis onto the basal plane is shown in Fig. 1.2 (Aoki, 1994). The hydroxyl ions lie at the corners of the projected basal plane and occur at equidistant intervals (0.344 nm) along the columns perpendicular to the basal plane and parallel to the c-axis. Six of the 10 calcium ions in the unit cell are associated with the HA in these columns, resulting in strong interactions among them.

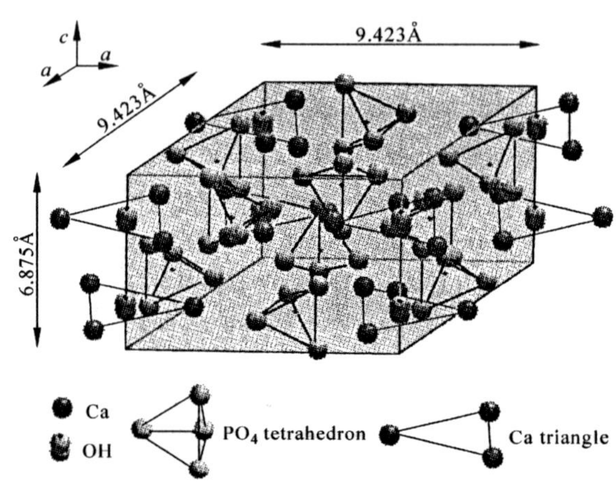

Figure 1.1 Crystal structure of hydroxyapatite

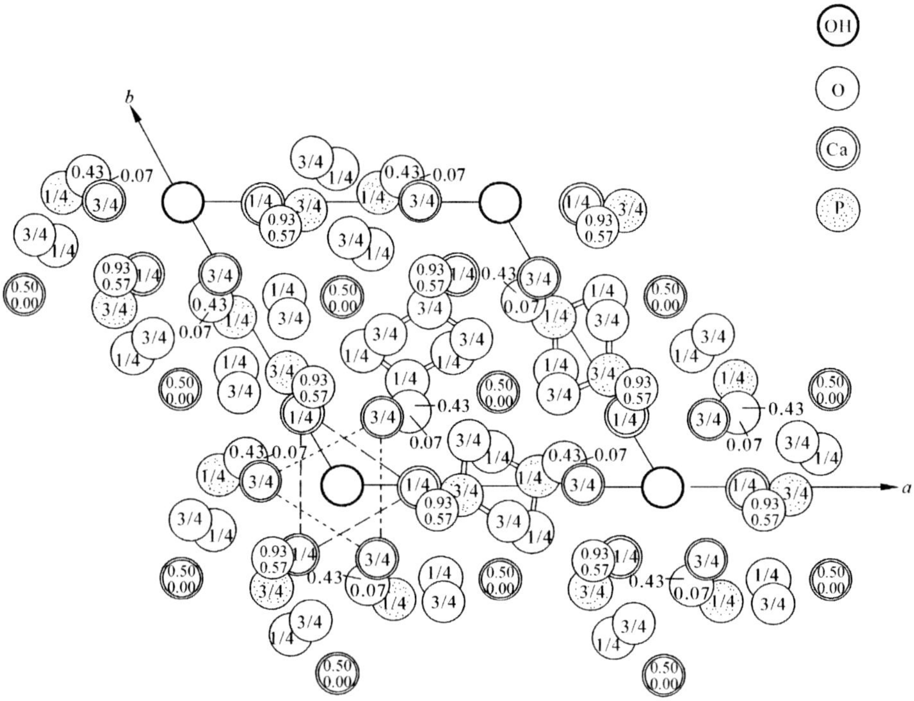

Figure 1.2 Structure of hydroxyapatite projected onto the basal plane

The crystallographic data of HA are summarized below:

Crystal system:	Hexagonal
Space group:	$P6_3/m$
Lattice constants:	$a = 0.9432$ nm, $c = 0.6881$ nm
Chemical unit number:	$Z = 1$
Molecular weight:	$M = 1004.8$

Substituted apatites such as $Ca_{10}(PO_4)_6F_2$ and $Ca_{10}(PO_4)_6Cl_2$ have similar structures, with F and Cl substituting of OH in its position.

The apatite structure is very hospitable in allowing the substitutions of many other ions. Ca, PO_4, and OH groups in apatite can be substituted. Carbonate (CO_3) can substitute either for the hydroxyl (OH) or the phosphate (PO_4) group, designated as Type A or Type B substitution, respectively. The substitutions cause morphological changes in precipitated apatite crystals as well as their properties. For example, CO_3 substituted apatite is more soluble than CO_3-free synthetic apatite.

1.2.2.2 Powder Synthesis

Hydroxyapatite in the particulate form can be produced by using a variety of methods. The characteristics of HA powders have significant effects on the subsequent products with HA being in the form of a dense or porous structure, in coatings, or in composites.

Wet Method

The wet method can be used for the mass production of crystalline HA or noncrystalline calcium phosphate powder. There are typically two types of process in the wet method: one involves the neutral reaction of acid and alkaline solutions; the other involves the reaction of calcium salts and phosphate salts. The chemical reactions are shown below:

(i) $$10Ca(OH)_2 + 6H_3PO_4 \rightarrow Ca_{10}(PO_4)_6(OH)_2 + 18H_2O \quad (1.2)$$

(ii) $$10CaCl_2 + 6Na_2HPO_4 + 2H_2O \rightarrow Ca_{10}(PO_4)_6(OH)_2 + 12NaCl + 8HCl \quad (1.3)$$

$$10Ca(NO_3)_2 + 6(NH_4)_2HPO_4 + 2H_2O \rightarrow Ca_{10}(PO_4)_6(OH)_2 + 12NH_4NO_3 + 8HNO_3 \quad (1.4)$$

The first synthetic HA powder was prepared by using the method developed by Tagai and Aoki (Tagai and Aoki, 1978). Production involved preparing a 200 ml suspension of $Ca(OH)_2$ and slowly adding 200 ml of 0.28 M H_3PO_4. The solution was maintained at a temperature of (50 ± 1)℃ and agitated with a magnetic stirrer for the duration of the addition process and for a subsequent 24 h after acid addition had ceased. The resulting precipitate was allowed to settle for a period of days, and some of the water was permitted to evaporate. After this, the powder was dried in an oven at 110℃.

Dry Method

The dry method is utilized to prepare crystalline HA by employing a solid-state reaction. For example, when brushite, $CaHPO \cdot 2H_2O$, and calcium carbonate are used as starting materials for the preparation of HA, the following continuous reaction occurs (>900℃):

$$6CaHPO \cdot 2H_2O + 4CaCO_3 \rightarrow Ca_{10}(PO_4)_6(OH)_2 + 4CO_2 + 14H_2O \quad (1.5)$$

HA synthesized by the dry method is very fine, and the crystallinity is excellent.

Hydrothermal Method

As mentioned above, HA prepared by the wet method in atmosphere consists of very small crystals with lattice defects. In order to obtain large, perfect, single crystals of HA, it is preferable to use the hydrothermal technique. Peroff and associates succeeded in growing an HA crystal to 0.3 mm under hydrothermal conditions of 300℃ and about 85 kg/cm^2, Mengeot and associates synthesized an HA single crystal of 7 mm × 3 mm × 3 mm, and Eyses and associates synthesized an 8 mm crystal (Aoki, 1994). The lattice constants of HA prepared by the hydrothermal method depend on the reaction temperature and pressure.

1.2.2.3 Manufacture of Dense Hydroxyapatite

HA can be prepared in a dense or macroporous object with pores being as large as 500 μm. Dense HA is described as having a maximum microporosity of 5% by volume with the micropores measuring about 1 μm in diameter and consisting of crystals with size exceeding 200 nm. Dense HA is generally sintered under specific conditions. Density up to 95% of the theoretical value is achievable. A flow chart of the fabrication process of dense HA by pressureless sintering, hot pressing, and hot isostatic pressing techniques is shown in Fig. 1.3.

Green Body Formation

This is the first step in producing dense HA stock or products. It involves forming HA powders into desired shapes. The precursor in the production of

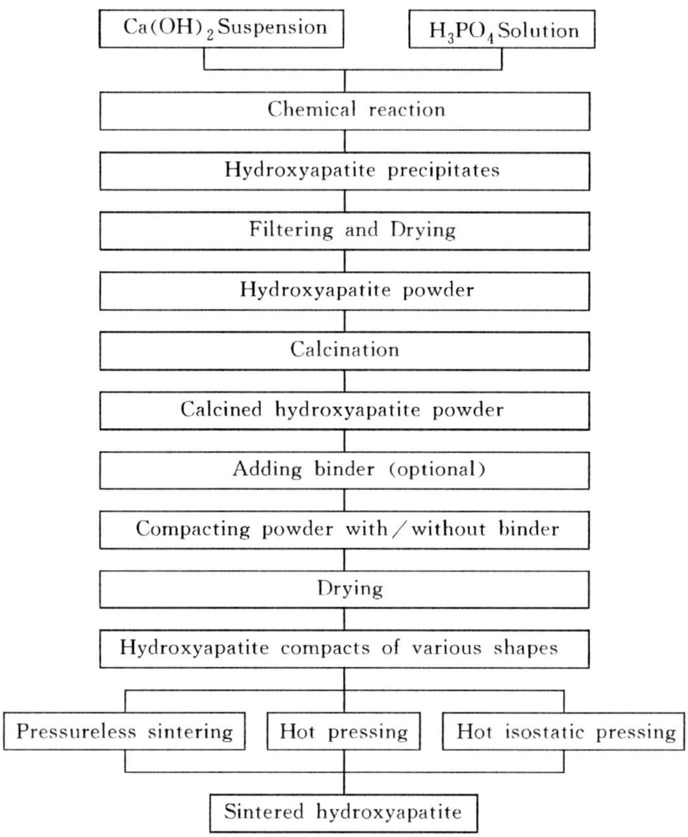

Figure 1.3 Fabrication of dense hydroxyapatite starting from powder synthesis

ceramic components is a low density and particulate (green) compact. This green compact is generally of low strength since particles are held together only by frictional / mechanical interlocking and low energy bonds such as van der Waals forces.

The primary goal of every compacting technique is to obtain homogeneity in all aspects, combined with the secondary goal of sufficiently high density of the green body. A variety of compacting techniques are used since the advantages and disadvantages differ per technique. They are divided into essentially dry consolidation (pressing), consolidation with doughs (injection molding or extrusion), and consolidation with slips (slip or tape casting). In pressing, there are two techniques that can be used: die pressing and isostatic pressing. Using die pressing, a certain amount of powder is put into a die and shaped by punches under a load. Since in this method the load is applied uniaxially, the method is also called (pseudo-) uniaxial pressing. Uniaxial pressing is the common method for compacting powders. During isostatic pressing, a powder batch is consolidated by isostatic pressure applied by a fluid on a reshaped compact protected with an impermeable cover. Technical details of isostatic pressing are given by James (James, 1983). The basic concept behind injection molding of ceramic parts is to put the numerous possibilities available in plastics technology to good use in ceramics. In ceramic processing, a thermoplastic material is mixed with ceramic powder. The process consists essentially of three steps: first, the filling of a relatively cold mold with the hot melt of the feedstock, i. e., a thermoplastic material mixed with the ceramic powder. Secondly, cooling and solidification of the melt, and, third, ejection of the formed product. After the injection molding process, the organic part has to be removed, usually, by burn-out. Details of ceramic powder injection molding are discussed elsewhere (German, 1990). For slip casting, various modifications exist. In all casting techniques, the use is made of a slip that is poured into a porous mold through which the solvent is removed either by capillary or by external pressure. Fairly complicated shapes can be produced (by using multipart molds) at relatively low cost, but the production speed is low.

Viscous plastic processing (VPP) has been in use for producing high-strength ceramics and was recently investigated for dense HA (Shaw et al., 1995). VPP operates by incorporating a high rate mixing step to break down agglomerates and simultaneously form a monolayer of polymer over each particle to reduce reagglomeration. This homogenization of the starting powder effectively removes potential crack initiation sites. Thus, the strength of ceramics processed through VPP is improved over those processed through conventional techniques such as pressing or slip casting. During VPP, the HA powder is loosely mixed with an organic binder and solvent to produce a crumb-like green mixture. This green mixture is then poured into the nip of a twin-roll mill, where it is calendered. After rolling, the material is pressed under a load between polyethylene sheets for two hours in order to remove any entrapped air. Disks are cut from the pressed sheet and then dried. For comparison, HA disks were also produced using uniaxial pressing. All disks were sintered using similar schedules. It was found that VPP resulted in increased densification for a given sintering temperature and led to a marked improvement in strength for sintered HA.

Densification

Densification of ceramic compacts can be carried out in various ways. Almost all of them are variations of sintering. Sintering brings the compacts to such a high temperature that sufficient mobility is present to release the excess surface energy of the powder, thereby joining the particles together. If diffusion takes place only in the solid state, the process is called solid-state sintering. If enhanced mobility is realized by a small amount of a liquid-phase, then the term liquid-phase sintering is used. The application of external pressure on the compact during sintering is called pressure sintering or hot pressing. The pressure can be applied uniaxially or isostatically.

The driving force for sintering is the reduction in the total free energy of the body. This process is often associated with a volume reduction, due to the misfit of the original particles, as the void volume is eliminated. The following criteria must be met in order that sintering occurs:

(1) A mechanism for material transport must exist.

(2) A source of energy is required to activate and sustain material transport.

There are four mechanisms by which sintering may take place:

(1) Vaporphase sintering – this is important in only a few material systems.

(2) Solid-state sintering – this occurs at a temperature below the melting point of any of the constituent phases and involves material transport by diffusion.

(3) Liquid-phase sintering – this requires the presence of a liquid at the sintering temperature and is the primary densification mechanism for most vitreous systems.

(4) Reaction sintering – this involves the heat treatment of a homogeneous mixture of two or more reactants, giving a dense structure formed by the products of the reaction.

Pressureless Sintering

Without external pressure, HA green bodies can be sintered in air at elevated temperatures, usually between 950℃ and 1300℃, in a conventional high-temperature furnace. In one early experiment, fine HA powder, mixed with 1 wt% cornstarch and a few drops of water, was pressed in a mold at a pressure of 60 – 80 MPa. The powder compacts were heated in air at 1150℃, 1200℃, 1250℃, and 1300℃, respectively, for 3 hours. α- and β-TCP phases were not detected in the sintered HA through X-ray powder diffraction analysis (Aoki, 1994).

Hot Pressing

Using a hot pressing apparatus, ceramic raw materials can be sintered at high temperatures while under high pressure at the same time. The apparatus may consist of a press and a furnace. Generally, the hot pressing technique is used for the fabrication of dense ceramic products at lower temperatures than the pressureless sintering technique. For HA, this procedure allows densification to take place at a much lower temperature (e. g., 900℃) which prevents the formation of other calcium phosphate phases such as α- and β-TCP and TTCP. It appeared that the temperature for starting sintering did not depend on pressure and that it was approximately 640℃. On the other hand, the completion of sintering greatly depended on pressure. By choice of suitable hot pressing conditions, transparent sintered HA could be obtained (Aoki, 1994).

Hot Isostatic Pressing

Hot isostatic pressing (HIPing) is a technique for producing ceramic bodies by gaseous pressures at elevated temperatures. By using this technique, ceramics with small grain sizes and few pores can be obtained at relatively low sintering temperatures. It was reported that HA powder was firstly uniaxially pressed at 80 MPa using a metal mold to obtain HA compacts. The HA compacts were then sintered at 1100℃ under 160 MPa of argon gas pressure for 1 h. Transparent sintered HA was thus obtained (Aoki, 1994). Greater density and higher strength of HA can be achieved using HIPing than using other densification techniques.

In summary, the production of dense HA consists of the following steps:

(1) preparing HA powder in-house or using commercially available HA powder;

(2) compacting the HA powder into a desired shape and size under high pressure;

(3) sintering the compact;

(4) using optional postsintering treatment to improve the properties of the HA product.

Pure HA has the theoretical composition of 39. 68 wt % Ca and 18. 45 wt% P and a Ca/P molar ratio of 1. 67. If the Ca/P molar ratio is 1. 67, only HA will be observed in the X-ray diffraction pattern and infrared spectrum. If the Ca/P molar ratio is lower than 1. 67, β-TCP, TTCP, and other phases will be present with the HA phase in the sintered material. If the Ca/P molar ratio is higher than 1. 67, CaO will be present with the HA phase. The purity, composition and the particle size of the apatite prepared before sintering and the sintering temperature and conditions (e. g., with or without water pressure present) affect the type and amount of other calcium phosphates (or compounds) in the sintered dense HA. They also have an influence on the microstructure of sintered HA.

1.2.2.4 Thermal Stability of Hydroxyapatite

It was reported that HA became unstable at 1000℃ and phase decomposition occurred after heat treating the spray dried powder at 1000℃ for 10 min. $Ca_3(PO_4)_2$ and CaO began to form, but the major phase was $Ca_{10}(PO_4)_6(OH)_2$. A longer thermal exposure at 1000℃ caused further decomposition of $Ca_{10}(PO_4)_6(OH)_2$ and the formation of other phases such as $Ca_4H_2(P_3O_{10})_2$. Despite this decomposition, the major phase was still HA. The following chemical reactions are proposed to describe the phase decomposition (Luo and Nieh, 1995):

$$Ca_{10}(PO_4)_6(OH)_2 \xrightarrow{1000℃/1.5\ h} 3Ca_3(PO_4)_2 + CaO + H_2O \quad (1.6)$$

$$2Ca_{10}(PO_4)_6(OH)_2 \xrightarrow{1000℃/1.5\ h} 3Ca(PO_3)_2 + Ca_4H_2(P_3O_{10})_2 + 13CaO \quad (1.7)$$

In another report, when heated to a high temperature, HA transformed into TCP and TTCP (Van Landuyt et al., 1995):

$$Ca_{10}(PO_4)_6(OH)_2 \rightarrow 2\ \alpha\text{-}Ca_3(PO_4)_2 + Ca_4O(PO_4)_2 + H_2O \quad (1.8)$$

A temperature of 1300℃ is usually considered the maximum sintering temperature because HA dissociates around this temperature into TCP and TTCP.

To maintain high thermal stability of HA during the whole manufacturing process, one way is to optimize the sintering condition, such as using high pressure or a steam atmosphere (Locardi and Pazzaglia, 1993). The other way of combating the thermal stability problem is to strictly control the properties of the starting powder. It is well known that there are very close relationships between the characteristics of starting ceramic powders and the microstructure and properties of a bulk material (Khor and Cheang, 1994; Best and Bonfield, 1994). *In vivo* studies showed that the biological performance and osteoblastic response of implants also depended on the chemical composition and physical properties of the original powder (Best et al., 1997).

The thermal stability of synthetic HA powders is a complicated issue. It was reported that when heated to a temperature higher than 900℃, partial dehydration may take place as follows (Fang et al., 1994):

$$Ca_{10}(PO_4)_6(OH)_2 \xrightarrow{900℃} Ca_{10}(PO_4)_6(OH)_{2-2x}O_x + xH_2O \quad (1.9)$$

With increasing temperature, the apatite structure changes to a new space group and TCP appears. Generally, β-TCP is more stable than α-TCP under 1200℃ and will be found first in HA which is subjected to elevated temperatures. Above 1200℃, β-TCP changes to α-TCP.

Vogel et al. studied the thermal stability of HA at a still higher temperature and reported that at 1300℃, HA would decompose as shown below (Vogel et al., 1996):

$$Ca_{10}(PO_4)_6(OH)_2 \xrightarrow{1300℃} 2\alpha\text{-}Ca_3(PO_4)_2 + Ca_4P_2O_9 + H_2O \quad (1.10)$$

This reaction is followed by

$$Ca_4P_2O_9 \rightarrow \alpha\text{-}Ca_3(PO_4)_2 + CaO \quad (1.11)$$

The effects of HA precipitation reaction parameters on the thermal stability of HA powder were recently investigated (Yang and Wang, 1999a). Experimental results showed that for HA powder, low reactant concentration and high reaction temperature resulted in higher thermal stability. Powders produced with 0.5 M reactant concentration and at 60℃ could withstand a temperature of up to 1350℃ at which apatite was still the dominant phase. Powders produced at other parameters decomposed at lower temperatures between 700℃ and 900℃. The decomposition products of all HA powders were initially β-TCP and finally α-TCP. Lattice parameter measurement of as-precipitated HA indicated that low reactant concentrations resulted in smaller lattice parameters.

1.2.2.5 Solubility of Hydroxyapatite

Hydroxyapatite is soluble in acidic solution, insoluble in alkaline solution, and slightly soluble in distilled water. Its solubility in distilled water increases with the addition of electrolytes. It was reported that crystalline HA is thermodynamically the most stable calcium phosphate at the pH, temperature and composition of physiological fluids. Amorphous calcium phosphate and β-TCP are less stable and more readily resorbed by the body (Correia et al., 1996; Brown et al., 1994). Moreover, HA resorbs very slowly compared with β-TCP. Even with β-TCP, biodegradation has been reported to be incomplete 40 months after grafting in human periodontal intrabony defects (Knabe et al., 1998).

The solubility of HA changes in the presence of amino acids, proteins, enzymes, and other organic compounds. However, the dissolution rate depends on the shape, porosity, crystal size, and crystallinity of HA implants. The solubility of sintered HA is very low. In subcutaneous tissue, the rate of solubility is 0.1 mg/year. HA reacts actively to proteins, lipids, and other inorganic and organic materials (Aoki, 1994).

1.2.2.6 Mechanical Properties of Dense Hydroxyapatite

The earliest studies on the sintering of dense HA were carried out by Jarcho et al. in 1976 (Jarcho et al., 1976). They produced a green compact using a variation of the slip casting technique and sintered the as-produced precipitation cake. They reported a Knoop hardness of 480, compressive strength of 917 MPa, and modulus of 30 GPa for specimens sintered between 1000℃ and 1200℃. Thomas et al. carried out diametrical compression tests on samples of "Durapatite" which were 5 mm in diameter (Thomas et al., 1980). They found that tensile strength ranged from 109 MPa to 121 MPa, increasing with an increase in the crosshead speed of the testing machine. According to Thomas, polycrystalline HA has an elastic modulus of from 40 GPa to 117 GPa.

Evidently, the mechanical properties of HA are affected by its density. The denser the material, the stronger. Denissen and de Groot used continuous hot pressing to produce specimens of 97% of the theoretical density of HA (Denissen and de Groot, 1980a). The Vickers hardness was measured to be 500 MPa and the tensile strength, measured using three point bend tests on polished specimens, was found to be 250 MPa. Denissen also made an HA of 99% of theoretical density after heating a powder compact to 1200℃. The tensile strength was found to be 390 MPa and Vickers hardness 450 MPa. de With made samples of 98% of theoretical density by using the hot isostatic pressing technique (de With, 1981). These samples had a fracture strength between 92 MPa and 103 MPa, Young's modulus of 116 GPa, and fracture toughness of 1.33 $MPa \cdot m^{1/2}$.

It is apparent that the properties of the HA powder and the compacting and sintering conditions have significant effects on the mechanical properties of dense HA. The presence of β-TCP in sintered HA causes a decrease in the fracture toughness of HA. Sintered dense HA is fragile. It is unsuitable for load-bearing applications in spite of its good biocompatibility and osteoconductivity.

1.2.2.7 Manufacture of Porous Hydroxyapatite

In producing porous HA, or porous calcium phosphate, two approaches are followed: one is to use living organisms to produce the desired products, i.e., from natural sources (Chiroff et al., 1977). Although the materials could be of composition similar to those of living tissue or even be produced by living organisms, they cannot be called bone graft because they do not comprise human or animal cells. The other approach is to use common synthetic methods. In most cases, making HA powders is the first step. The HA powders are then used in a number of ways to produce porous HA.

Hydrothermal Conversion of Corals

The use of corals from Goniopora, Porites, Favites, etc., has been investigated (Bucholz et al., 1987; Guillemin et al., 1987; Holmes, 1979). These coral specimens consist of 99% calcium carbonate in the form of aragonite and 1% organic material. After being ultrasonically cleaned, corals were heated at elevated temperatures and under pressure in the presence of aqueous phosphate solutions. This caused a replacement of phosphate ions for the carbonate ions and changed the crystal structure to calcium HA. Typically, the X-ray diffraction pattern of this calcium phosphate was predominantly HA. The chemical reaction for the process is shown below (Roy and Linnehan, 1974):

$$CaCO_3 + (NH_3)HPO_4 \xrightarrow{275℃, 12000\ psi} Ca_{10}(PO_4)(OH)_2 \quad (1.12)$$

The porosity of these corals has advantages and disadvantages for implantation purposes. In the early investigation of corals for use as bone graft substitutes, Chiroff et al. observed the "remarkable" uniformity of pore size and complete interconnection of the pores (Chiroff et al., 1977). Further studies have shown that certain corals have uniform porosity in the range of 500 – 60 μm in diameter, which is known to be advantageous for osseointegration. In particular, the Porites and Goniopora corals have an open pore structure (interconnecting pores), which is believed to allow for rapid vascularization and osseointegration of the implant. One disadvantage of using corals is that the porosity varies only slightly from 46% to 48%. This level of porosity causes coral to be weaker than cancellous bone. The fixed amount of porosity may also be biologically undesirable in certain instances. Another disadvantage of the substituted corals is the variation in the final chemical composition of the ceramic. It is known that ions such as magnesium, sodium, chloride, and fluoride, which are present in seawater and become incorporated into the coral, affect the sintering parameters of these calcium phosphate implants and may have an effect on the biological response. The difficulty in obtaining sufficient starting material from commercial sources is also a disadvantage of this method.

Porous Hydroxyapatite from Bovine Bone

A porous ceramic structure was developed using cow bone as the raw material (Uenu, 1987). Both cancellous and cortical portions of cow bone were used. Soft tissues such as muscles and tendons were removed from the frozen bone. The bones were then cut into 10 cm^3 pieces. These pieces were treated for deproteinization by boiling in water for 10 h before their immersion in a 1% sodium hydroxide plus 1% hydrogen peroxide solution for 1 h. They were then washed under flowing water and, after drying, they were sintered in an electric furnace. During sintering, organic substances were completely eliminated and crystallization of bone minerals occurred. The sintered pieces were shaped into small blocks, 3 – 10 mm^3 in size, and sterilized with ethylene oxide gas or in an autoclave. This porous material was said to have superior biological properties and is expected to be a substitute for autograft.

Polymer Porosifier Method

Liu et al. used polyvinyl butyral (PVB) particles as a porosifier to make porous HA ceramic material (Liu et al., 1996). PVB particles with average sizes of

0.093 mm, 0.188 mm, and 0.42 mm were used. Powder mixtures containing HA granules and 42% – 61% by volume of PVB particles were prepared. The mixed powders were die-pressed uniaxially at pressures of 27 MPa and 55 MPa to form rectangular blocks. The blocks were then heat treated to 500℃ at a heating rate of 0.5℃/min to drive off the PVB and other volatiles, followed by an increase to 1200℃ for 2 h to 48 h for densification purposes. Experimental results showed that HA ceramics with controlled pore characteristics such as pore volume fraction, pore size and pore structures were achievable. This provides the possibility of designing HA with diverse porosity simulating that of natural bone.

Polymer Foaming Method

Polymer foams coated with ceramic powders can produce open-cell structures, and this method applies to a wide range of ceramics. First, an appropriate polymeric sponge is selected. Sponges with different characteristics can be commercially obtained. Two aspects are critical to the selection of sponges: one is the pore size, which determines the final pore size of the ceramic, and the other is the type of polymer. Polyurethane is the most attractive polymer because of its low softening temperature and ease of burn-off which minimize thermal stresses that may fracture the unsintered ceramic. The sponges are coated with HA powder slurry, then dried and sintered. In this method, sintering aids and binding agents were used (Liu and Brown, 1992; Fabbri et al., 1995; Sherman et al., 1991; Lange and Miller, 1987; Brezny and Green, 1989).

H_2O_2 Foaming Method

Rejda et al. mixed an HA slurry with H_2O_2 to make porous HA. The porous bioceramic had pore sizes in the ranges of 0.5 – 1.5 μm (micropore) and 150 – 250 μm (macropore) that were controllable by varying the hydrogen peroxide concentration (Rejda et al., 1977). A similar experiment was conducted to produce macro-porous HA (Klein et al., 1989).

Solid Reaction Method

It is known that $CaCO_3$ can be used as a gas-forming agent to react with dicalcium phosphate at 1000℃:

$$6CaHPO_4 + 4CaCO_3 \rightarrow Ca_{10}(PO_4)_6(OH)_2 + 2H_2O\uparrow + 4CO_2\uparrow \quad (1.13)$$

This reaction was also mentioned by Aoki as a dry method for HA powder preparation (Aoki, 1994). Later, Arita et al. made use of this reaction to fabricate porous HA (Arita et al., 1995). The pores can be created by the evolution of gas phase CO_2 and H_2O during reaction, and the size of the pores can be controlled by the particle size of the starting $CaCO_3$. By mixing different volume fractions of precalcined HA powders into the $CaHPO_4$ + $CaCO_3$ reactants, various levels of porosity are achievable.

Coating Porous Ceramics with Hydroxyapatite

Porous HA is known to have low fracture strength and poor fatigue resistance. In order to seek better ways of processing HA so that both mechanical properties and porosity could be optimized, a thin layer of HA was coated onto the inner surfaces of pores in reticulated alumina (Jiang and Shi, 1998; Shi et al., 1999). Using this method, the basic requirements for bone implants may be met: (1) a high porosity level due to the nature of the reticulated alumina for an organization of vascular canals; (2) compatible mechanical properties with the bone structure, and (3) a high bioactivity induced by the thin layer of HA for bone ingrowth. HA can be uniformly coated onto both inner and outer surfaces of reticulated alumina substrates. The *in vitro* bioactivity of HA coatings was found to be strongly affected by the crystallinity of the HA.

1.2.2.8 Mechanical Properties of Porous Hydroxyapatite

HA, as a ceramic material, is brittle and vulnerable to tensile stress. With porous HA, the situation is worse because many pores, being large or perhaps pore clusters, act as critical flaws in the structure. When there are no sharp corners in these pores, the influence of pores as failure inducing flaws must be taken into account. Generally, the mechanical properties of porous HA are poor. Table 1.5 gives a comparison of the mechanical properties of porous HA, dense HA, and cortical and cancellous bones (LeGeros, 1991; Aoki, 1994).

Table 1.5 Mechanical properties of hydroxyapatite compared with human bone

Material	Compressive Strength /MPa	Tensile Strength /MPa	Modulus /GPa
Cortical bone	137.8	68.9	13.8
Cancellous bone	41.4	3.5	3
Porous HA	6.9 - 68.9	2.48	/
Dense HA	206.7 - 895.7	68.9 - 192.9	34.5 - 103.4

It was shown recently that the biaxial flexural strength of porous HA follows an exponential function of total porosity, as illustrated in Fig. 1.4 (Yang and Wang, 1999b). For all uniaxially pressed, sintered HA with porosity from 14.2% to 52.2%, the biaxial strength varied from 38 MPa to 5.76 MPa. Porosity has the greatest influence on mechanical properties, while pore size, pore shape, and the interaction of pore size and pore shape also have significant effects.

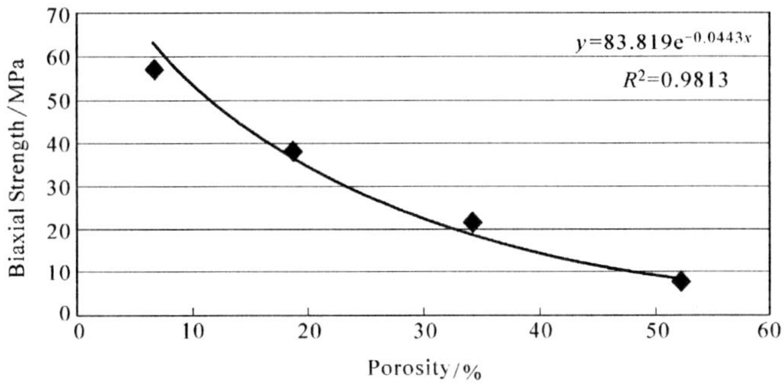

Figure 1.4 Influence of porosity on the biaxial strength of porous hydroxyapatite

1.2.2.9 Biological Properties of Hydroxyapatite

Biocompatibility

The use of HA as an implant material should be, in the first place, attributed to its good biocompatibility which probably stems from the perceived advantage of using a material similar to the mineral phase in bone and teeth for replacing these tissues.

The biocompatibility of HA has been widely studied (Denissen et al., 1980b; de Lange et al., 1987; Eggli et al., 1988; Maxian and Zawadsky, 1993; Hing et al., 1996; Best et al., 1997). All the systems investigated show good biocompatibility. Klein et al. emphasized that HA particularly displayed an attractive profile that features its excellent biocompatibility (Klein et al., 1989). de Lange et al. studied the bone – HA interface and found that the presence of HA did not disturb the normal process of bone tissue maintenance (de Lange et al., 1987). The host response to porous HA ceramics was claimed to be completely of osteoconduction (Liu, 1996). This means that HA permits bony ingrowth and acts as a scaffold in bony tissue sites. Porous HA, with a pore size greater than 100 μm, showed significant ingrowth of natural bone when these implants were in contact with host bone tissue. Hing et al. put three groups of porous HA specimens with mean apparent densities of 61 kg/m^3, 90 kg/m^3, and 1200 kg/m^3 into cancellous bone of rabbits for a period of 5 weeks (Hing et al., 1995). Low density specimens exhibited behaviour similar to trabecular bone while the high-density material, which had a higher ultimate compressive stress than the cancellous bone, still behaved as a brittle material. HA specimens were reported to increase their compressive strength after implantation because bone ingrowth effectively reinforced the implant and improved its mechanical properties over time. In orthopedics, HA is primarily used as a coating for prostheses. The HA coating was claimed to act as a stimulus for bone growth (Geesink, 1991). These coated prostheses were said to be more "forgiving" in the sense that gaps left between the prostheses and the bone became filled with bone which was stimulated to grow by the HA-coated implant.

One of the implant – tissue interactions that requires consideration is biodegradation. A biodegradable material is one that is dissolved without provoking any adverse tissue reaction after implantation. It would be ideal that, having served its assigned purpose, an implant would gradually degrade and be replaced by normal bone tissue to achieve perfect recovery. Degradability is perhaps desired of an HA implant because no hardware removal will be necessary later for this type of biomaterial. However, whether HA is degradable or not is still controversial. In many cases, HA is not considered degradable. Some researchers showed that HA was not degradable even after 3.5 years of implantation (Hoogendoorn et al., 1987). On the other hand, it was reported that high resorption rates of HA samples could be observed in direct contact with soft tissue (Janicke and Wagner, 1988). It was found that degradation of HA did occur, but it stopped after osteoingrowth (Kasperk and Ewers, 1987). It was shown recently that when put into a wet environment, the strength of HA would decrease by 9% (Ito et al., 1996). In fact, the biodegradability of HA is not simply decided by the material itself. It has been pointed out that the factors influencing biodegradation or bioresorption of HA are density, pore characteristics, composition (impurities), crystallinity, and solubility (Toth et al., 1995).

Bioactivity

Another attraction for using HA is its bioactivity. The concept of bioactivity can be considered with respect to bioactive glasses via this hypothesis: the biocompatibility of an implant material is optimal if the material elicits the formation of normal tissue at its surface and, in addition, if it establishes an interface capable of supporting the loads which normally occur at the site of implantation (Ducheyne, 1987). Bioactive materials can form a chemical bond with osseous tissue. Bioactivity has been demonstrated by active osteoblasts on the surface of implants for most of the calcium phosphate ceramics. Bone forms a very tight bond with HA. Many researchers have shown convincing evidence that biological apatite on synthetic HA in a collagenous matrix on synthetic HA is in the same crystalline orientation as the synthetic material. This strong bonding is particularly evident in HA-coated implants when failure analysis is performed on the implant – bone interface.

The bioactivity of HA may be illustrated by the healing process at the implant site. During the wound healing process associated with implantation of the material, angiogenesis takes place and a capillary blood supply develops. Healing fibroblasts attach to the interface and form an appositional fibrous matrix. If vascular penetration into the implant can occur, dependent upon composition and structure, some of the multipotential mesenchymal cells differentiate into osteoblasts, and endosteal membrane develops. This membrane actively produces dense fibrous osteoid that becomes calcified. The orientation of the osteoid is perivascular and appositional to the implant surface, thus primary vascular bone develops from the endosteal membrane activity. Osseointegration occurs, and the implant becomes part of the bone at the site.

Pore Effect on Bone Ingrowth

It has been known for about 20 years that tissue can grow into implanted porous structures, the nature of the tissue being controlled by the nature of the porosity (Williams, 1987). If the pores are interconnecting and the minimum pore size is in the region of 150 – 400 μm diameter, then bone will grow into the pores, irrespective of the nature of the material, provided it is reasonably stable chemically (Piotrowski et al., 1975). This is a pore effect, and this effect is simply affected by pore properties. Therefore, porous metal, Al_2O_3 and silica, like porous HA, all can promote bone ingrowth into their pores.

To understand the pore effect, the physiological response of tissue needs to be considered. Porous HA is used here as an example. As the macrophages, angioblasts, and fibroblasts proliferate, they are able to invade the pores. When phagocytosis occurs, resorption cavities enlarge, and angioblasts can become endothelial cells forming capillaries and extend the blood supply through interconnecting channels. Healing fibroblasts produce a fibrous collagen matrix. This matrix increases in density and, if the macrophages, histiocytes, and giant cells can successfully resorb the porous material to increase the areas into which the capillaries can penetrate, the fibrous matrix will extend itself. Meanwhile, at the ceramic – bone interface, in a favorable environment with an adequate blood supply and various growth factors, some of the multipotential mesenchymal cells are induced to differentiate into osteoblasts, and an endosteal osteogenic membrane appears. The osteoblasts secrete a fibrous Type I collagen matrix called osteoid, laid down appositionally on the surface of existing bone and also on the scaffold walls of the porous structure. Osteoid is the precursor matrix that, when calcified, becomes bone. Therefore, in this process, bone is said to "grow into" the pores of an HA implant.

1.2.2.10 Clinical Applications

The principal objective of using HA implants, dense or porous, is to obtain a well osseointegrated bone-to-implant bridge with sufficient strength to provide normal structure and function. Bulk HA has been used with success in a variety of applications (Hench, 1991). It has been used by some researchers with a certain degree of success for the restoration of atrophic mandibular ridges in animal studies as well as human clinical trials. It has been used as a filler for periodontal defects, alveolar ridge augmentation, and maxillofacial reconstruction. When used as a porous ceramic, results have been very good. Porous HA is mainly used as cancellous bone substitute or specific filler due to its comparatively low mechanical properties. HA used as carriers for antibiotics has also been reported. In addition to being used on its own, HA is also used as a constituent of various composites, as coatings on metallic stems of hip prostheses, and as the second phase in PMMA bone cement. Table 1.6 lists the application history of HA implants.

Table 1.6 Applications of hydroxyapatite

Year	Event	Researcher
1971	Suggested use of HA as bone and tooth implant	More et al., 1971
1974	Suggested use of HA as orthopedic implant	Hubbard, 1974
1975	Suggested use of HA for periodontal treatment	Nerf et al., 1975
1976	Development of processes for preparing dense polycrystalline HA	Jarcho, et al., 1976
1977	Use of HA as loaded tooth roots in animals	Jarcho, 1977
1978	Bonding of bone to HA	Boyne and Shapton, 1978
1979	Clinical application of HA dental root implants in Europe	Denissen and de Groot, 1980c
1980	Role of macro- and microporosity in resorption rate and biological response of HA identified	de Groot, 1980
1980	Clinical trials of HA powder in alveolar ridge augmentation	Kent et al., 1980
1980	Use of HA powder to stimulate bone growth in porous metal implants	Ducheyne et al., 1980
1981	Clinical application of HA loaded tooth implants in Japan	Ogiso et al., 1981
1981	Clinical use of dense HA in middle ear canal wall prostheses	Grote et al., 1981

1.2.3 Tricalcium Phosphate

The chemical formula of tricalcium phosphate (TCP) is $Ca_3(PO_4)_2$. TCP has four polymorphs: α, β, γ, and super-α. The γ polymorph is a high-pressure phase, and the super-α polymorph is observed at temperatures above approximately 1500℃ (Gibson et al., 1996). Therefore, the most frequently observed TCP polymorphs in the field of bioceramics are α and β-TCP.

α-TCP crystal is in the monoclinic space group $P2_1/a$ with lattice parameters $a = 1.2887$ nm, $b = 2.7280$ nm, $c = 1.5219$ nm, and $\beta = 126.20°$. There are 24 formula units per unit cell. There is a prominent approximate subcell with a b axis parameter of $b/3$ (0.909 nm) that contains eight formula units. The structure comprises columns of Ca^{2+} and PO_4^{3-} ions parallel to the c axis, arranged as shown in Fig. 1.5 (Elliott, 1994).

β-TCP has the rhombohedral space group R3c with unit cell $a = 1.0439$ nm, $c = 3.7375$ nm (hexagonal setting) with 21 formula units per hexagonal unit

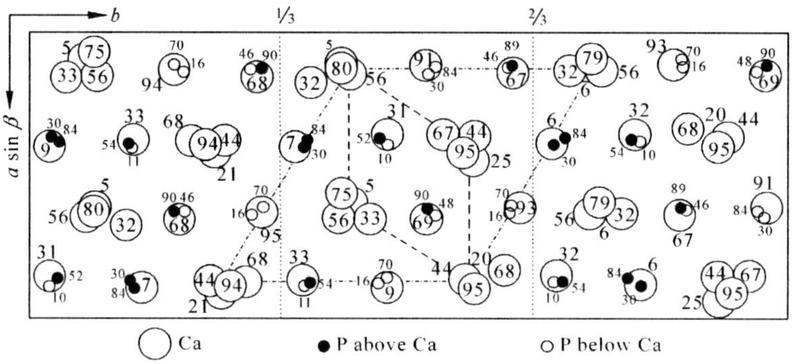

Figure 1.5 Projection of α-TCP structure onto the (001) plane (only the calcium and phosphorus atoms are shown)

cell. The structure of β-TCP has been determined and described in terms of a distortion from the $Ba_3(VO_4)_2$ structure. The β-TCP structure is similar to $Ba_3(VO_4)_2$ but has three fewer formula units per hexagonal unit cell, as shown in Figs. 1.6 and 1.7 (Elliott, 1994).

β-TCP is stable up to 1125℃. But above this temperature and up to 1430℃, α-TCP becomes the stable phase. Super-α-TCP forms between 1430℃ and the melting point at 1756℃. The phase diagram for the CaO – P_2O_5 system is shown in Fig. 1.8 (Elliott, 1994).

The dissolution rate of TCP was investigated by various researchers (Black and Hastings, 1998). Jarcho compared the relative dissolution rates of dense HA and TCP. The dissolution rate of TCP was 12.3 times higher than that of HA in buffered lactic acid solution (0.4 M, pH 5.2) and was 22.3 times higher than that of HA in buffered ethylene diamine tetraacetic acid (EDTA) solution (0.05 M, pH 8.2). Klein et al. (1989) carried out dissolution tests on HA and β-TCP with various degrees of porosity in buffered lactic acid solution. The dissolution rate of β-TCP was three times higher than that of HA. Ducheyne et al. (1980) compared the dissolution rate of six calcium phosphates in calcium and phosphate-free solution at pH 7.3. The dissolution rate increased in the following order:

$$\text{HA} < \beta\text{-TCP} < \alpha\text{-TCP} < \text{TTCP} \qquad (1.14)$$

Compared to β-TCP, α-TCP has a lower density and a higher free energy of formation and is therefore expected to be more reactive (Bohner and Lemaître, 1996). α-TCP is fairly reactive in aqueous systems and can be hydrolyzed to mixtures of DCPD, calcium-deficient oxyhydroxyapatite, and other calcium phosphates in varying proportions, depending upon the conditions. β-TCP does not form in aqueous systems. Some evidence showed that β-TCP is always more soluble than oxyhydroxyapatite, but above pH = 6, it is less soluble than other

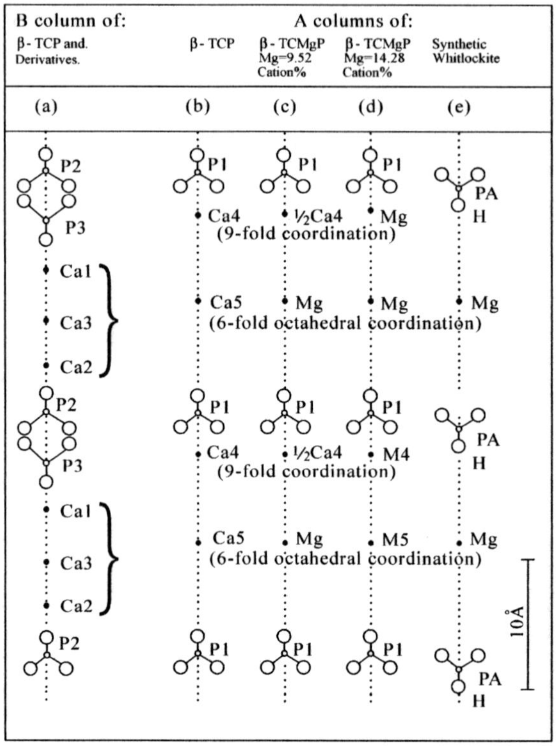

Figure 1.6 Configuration of A and B columns (length shown is *c* axis parameter of the hexagonal unit cell and additional PO_4 to show continuity of the columns)

(a) depict all the columns in $Ba_3(VO_4)_2$ and B columns in β-TCP and related compounds

(b) to (e) depict A columns in

(b) β-TCP;

(c) β-TCa, MgP ($Ca_{18}Mg_2(Ca, \square)(PO_4)_{14}$), all the Mg^{2+} ions have been put in the Ca(5), but structure analysis shows that some of them are in the Ca(4) sites;

(d) β-TCa, MgP ($Ca_{18}Mg_2(Mg, \square)(PO_4)_{14}$);

(e) whitlockite ($Ca_{18}Mg_2H_2(PO_4)_{14}$)

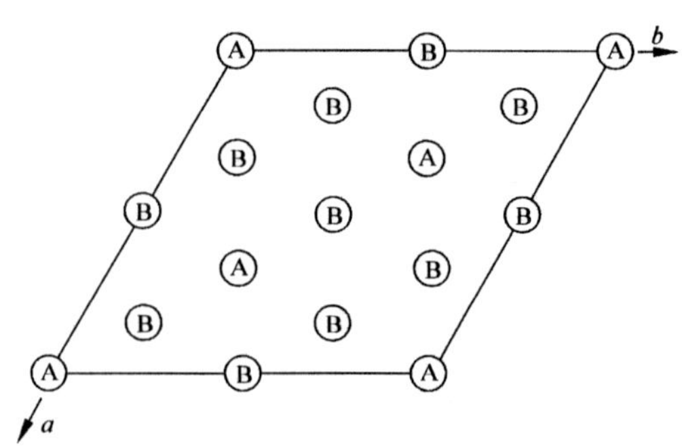

Figure 1.7 Arrangement of A and B columns in the hexagonal unit cell of β-TCP and its derivatives. (The *c* axis is out of the plane of the diagram.)

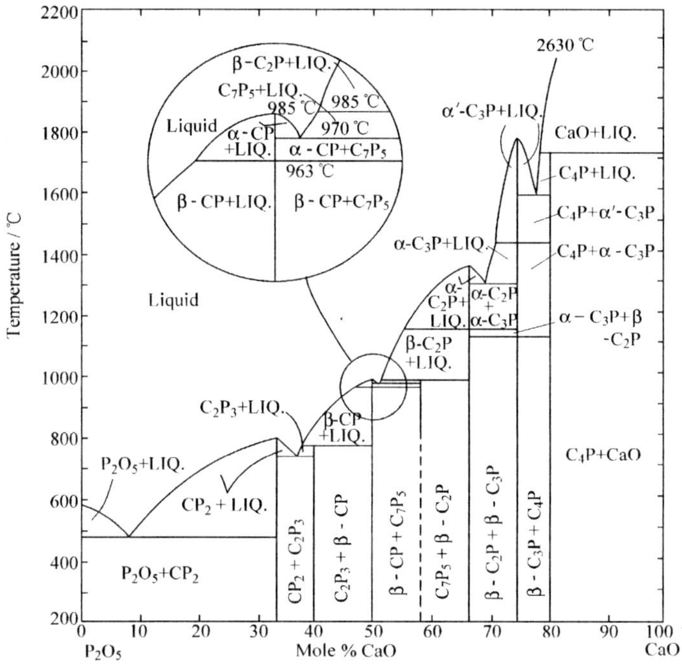

Figure 1. 8 Equilibrium phase diagram for the CaO-P_2O_5 system (C = CaO, P = P_2O_5)

calcium phosphates. Moreover, the solubility of β-TCP decreases with an increase in temperature (Elliott, 1994).

The mechanical properties of TCP are rarely available, but it was reported that TCP has slightly higher fracture toughness than HA (Santos et al., 1996). Fully dense and translucent β-TCP totally free from α modification can attain a maximum strength value of 120 MPa (Tampieri et al., 1997).

Most of the reports on TCP have concluded that TCP is biodegradable, although there are some differences, depending on the characteristics of the materials used (Black and Hastings, 1998). Due in part to its crystalline structure, the biodegradation rate of TCP has been shown to be much greater than that of HA. The exact mechanism for its biodegradation remains unclear. TCP has an important role as a resorbable bioceramic due to its high solubility and bioactivity. β-TCP has been accepted and used as a biocompatible, resorbable material for bone repair in the form of ceramic blocks, granules, and calcium phosphate cements (Bohner and Lemaître, 1996; Lemaître and Maréchalle, 1996).

Biodegradation of calcium phosphate ceramics is caused by the following factors:

(1) physiological dissolution, which depends on the solubility product of the material and local pH of its environment;

(2) physical disintegration into small particles as a result of preferential chemical attack at grain boundaries;

(3) biological factors, such as phagocytosis, which causes a decrease in local pH value.

For calcium phosphate ceramics, the rate of biodegradation increases as

(1) surface area increases (powders > porous solid > dense solid).

(2) crystallinity decreases.

(3) crystal perfection decreases.

(4) crystal and grain size decrease.

(5) there are ionic substitutions of CO_3^{2-}, Mg^{2+}, and Sr^{2+} in HA.

Factors that tend to decrease the rate of biodegradation include

(1) F^- substitution in HA,

(2) Mg^{2+} substitution in β-TCP,

(3) low β-TCP/HA ratios in biphasic calcium phosphates.

1.2.4 Other Calcium Phosphate Compounds

Biphasic calcium phosphate (BCP) which contains 60% HA and 40% TCP has been investigated for vertebral bone fusion and filling long bone defects (Daculsi et al., 1989; Daculsi et al., 1990). It was shown that bone formation and calcification appeared at the expense of BCP during the first few weeks of implantation. BCP promoted bone ingrowth at the implant – host tissue interface. BCP was said to be more efficient than pure HA for bone growth and bone substitution. BCP coating also provided a more efficient coating material than HA for accelerated interface attachment (Delecrin et al., 1991). Recently, macroporous BCP (70% porosity and 300 – 600 μm pore size) was investigated for cancellous and cortical bone substitution (Daculsi et al., 1999). It was found that the bone ingrowth during the first year in both human and animal was limited to 2.5 mm from the surface of the implants. In the cortical bone site, 50% of the ceramic was replaced by cortical lamellar bone after 1 year. In cancellous bone, large differences appeared between animal species and human due to differences in cell recruitment.

The growth and dissolution of brushite (DCPD) play important roles in the biological mineralization process as well as the setting of a variety of calcium phosphate cements for orthopedic and dental uses (Royer et al., 1992; Van Landuyt et al., 1998). Early reports on DCPD suggested that DCPD was one of the precursors of HA. DCPD can gradually transform to HA crystals under physiological conditions. Therefore, the dissolution kinetics of DCPD followed by precipitation of HA has drawn the attention of researchers in recent years. DCPD single crystals were produced recently using the gel technique (Kumar and Wang, 2000a, 2000b). In this technique, sodium silicate solution was used as the growth medium. Different inner and outer reactants were used for growth trials. DCPD needles of nearly 40 mm in length were obtained in a period of 45 days. As determined by nanoindentation tests, the crystals had average modulus and hardness values of 22.5 GPa and 1.75 GPa, respectively.

1.3 Bioactive Glasses and Glass-Ceramics

It was discovered by Hench and co-workers in 1969 that bone can bond chemically to certain glass compositions (Hench et al., 1972). This group of glasses has become known as bioactive glasses based on the following definition: "A bioactive material is one that elicits a specific biological response at the interface of the material which results in the formation of a bond between the tissues and the material." One aspect that makes bioactive glasses different from other bioactive ceramics and glass-ceramics is the possibility of controlling a range of chemical properties and rate of bonding to tissues. The most reactive glass compositions develop a stable, bonded interface with soft tissues.

By heat treatment, a glass can be converted into glass – crystal composites containing various types of crystalline phases with controlled sizes and contents. The resultant glass-ceramic can exhibit properties superior to the parent glass and to sintered crystalline ceramics. In the early 1980s, Kokubo and co-workers started the attempt to produce glass-ceramics that contain apatite and β-wollastonite (CaO · SiO) for bone substitution (Kokubo et al., 1982). The result is the development of a bioactive material that possesses excellent mechanical properties and can be used for a variety of clinical applications.

Currently, there are various glasses and glass-ceramics available for clinical applications (Table 1.7) (Hench, 1996). A common characteristic of bioactive glasses and glass-ceramics is the time-dependent, kinetic modification of the surface that occurs upon implantation. The surface forms a biologically active, carbonated apatite layer that provides the bonding interface with tissues. This adherent interface with tissues resists substantial mechanical forces. In many cases, the interfacial strength of adhesion is equivalent to or greater than the cohesive strength of the implant material or the tissue bonded to the bioactive implant.

1.3.1 Bioglass®

1.3.1.1 Composition

Bioglass® is a family of bioactive glasses that contain SiO_2, Na_2O, CaO, and P_2O_5 in specific proportions. These glasses differ from traditional soda – lime – silica glasses in three key compositional features:

(1) less than 60 mol% SiO_2,

(2) high Na_2O and CaO content,

(3) a high CaO/P_2O_5 ratio.

Table 1.7 Composition of bioactive glasses and glass-ceramics (in weight percentage)

Component	Glass and Glass-ceramic									
	45S5 Bioglass	45S5F Bioglass	45S5.4F Bioglass	40S5B5 Bioglass	52S4.6 Bioglass	55S4.3 Bioglass	KGC Ceravital	KGS Ceravital	KGy213 Ceravital	A-W GC
SiO_2	45	45	45	40	52	55	46.2	46	38	34.2
P_2O_5	6	6	6	6	6	6				16.3
CaO	24.5	12.25	14.7	24.5	21	19.5	20.2	33	31	44.9
$Ca(PO_3)_2$							25.5	16	13.5	
CaF_2		12.25	9.8							0.5
MgO							2.9			4.6
MgF_2										
Na_2O	24.5	24.5	24.5	24.5	21	19.5	4.8	5	4	
K_2O							0.4			
Al_2O_3									7	
B_2O_3				5						
Ta_2O_5/TiO_2									6.5	

These features make the glass surface highly reactive when it is exposed to an aqueous medium. Many bioactive silica glasses are based upon the formula called 45S5, signifying 45 wt% of SiO_2 and a 5 : 1 ratio of CaO to P_2O_5. Glasses with lower CaO to P_2O_5 ratios do not bond to bone (Hench, 1996). Hench and co-workers studied a series of glasses in the SiO_2 – Na_2O – CaO – P_2O_5 system with a constant 6 wt% P_2O_5 content. The addition of as little as 3 wt% of Al_2O_3 to the 45S5 formula prevents bone-bonding (Hench and Paschall, 1973; Hench and Clark, 1982; Greenspan and Hench, 1976). The experimental results are summarized in the ternary SiO_2 – Na_2O – CaO diagram shown in Fig. 1.9 (Hench, 1996). It gives compositional boundaries for bioactive glasses.

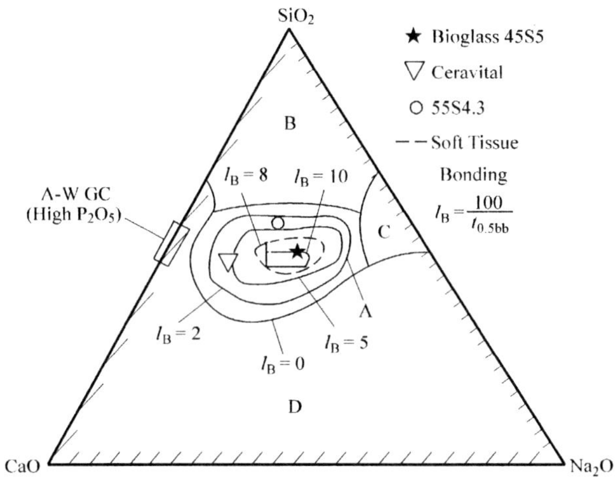

Figure 1.9 Compositional dependence of bone bonding and soft tissue bonding of bioactive glasses and glass-ceramics (All compositions in Region A have a constant 6 wt% of P_2O_5. I_B is index of bioactivity.)

In region A, the glasses are bioactive and bond to bone. In the middle of this area, a smaller region is indicated (broken line) within which soft tissue bonding also occurs. Glasses in region B behave as nearlyinert materials and are encapsulated by nonadherent fibrous tissue when implanted. Compositions in region C are resorbed within 10 – 30 days in tissue. In region D, the compositions are not technically practical. The boundary between region A and C depends upon the ratio of the surface area of the glass to the effective solution volume of the tissue, as well as the glass composition. Fine glass powders resorb more quickly than bulk implants.

1.3.1.2 Processing

Bioglass® is produced by the conventional glass-making method. Contamination of the glass must be avoided in order to retain the chemical reactivity of the material. Purity of the raw material has to be assured, and analytical grade compounds are typically used. Appropriate amounts of SiO_2, $CaCO_3$, Na_2CO_3, and $Ca_2P_2O_7$ are mixed and melted to produce Bioglass® 45S5. Weighing, mixing, melting, homogenizing, and forming of the glass must be conducted without introducing impurities or losing volatile constituents such as Na_2O and P_2O_5. Only platinum or platinum alloy crucibles or a glass melter may be used in order to avoid contamination of the melt.

Melting is usually performed in the range of 1300 – 1450℃, depending on glass composition. Bulk Bioglass® can be formed by casting or injection molding in graphite or steel molds. Annealing, conducted at 450 – 550℃, is crucial because of the high coefficient of thermal expansion of these glasses. If a granulated or powdered material is required, the melt can be rapidly quenched in water or air before grinding and sieving into the desired particle size. Bioglass® is soft compared to other glasses, and final shapes can be easily made by machining. Diamond cutting tools are preferred for machining. Dry grinding is also possible. Bulk or particulate Bioglass® must be stored in dry conditions as it easily absorbs water and reacts with it.

1.3.1.3 Properties

The primary disadvantage of Bioglass® is its mechanical weakness and low fracture toughness due to an amorphous two-dimensional glass network. The bending strength of most Bioglass® compositions is in the range of 40 – 60 MPa, which is not suitable for load-bearing applications. Young's modulus of Bioglass® is 30 – 35 GPa, which is close to that of cortical bone. Modulus matching of Bioglass® with cortical bone can be a distinctive advantage. Despite the inherent weakness, Bioglass® can be used for buried implants, in slightly-loaded or compressively loaded devices, in the form of powders, or as the bioactive phase in composites.

1.3.1.4 Surface Reaction and Tissue Bonding

The basis of the bone bonding property of bioactive glasses is the chemical reactivity of the glass in body fluids. The surface chemical reactions result in the formation of an hydroxycarbonate apatite (HCA) layer to which bone can bond. Bonding occurs due to a sequence of reactions. Upon immersion of a bioactive glass in an aqueous solution, three general processes occur: leaching, dissolution, and precipitation.

With a bioactive glass, the reactions on the implant side of the interface are (Hench and Clark, 1978; Bunker et al., 1988; Clark et al., 1976, 1990; Ogino et al., 1980)

Stage 1 Leaching and formation of silanols (SiOH),

Stage 2 Loss of soluble silica and formation of silanols,

Stage 3 Polycondensation of silanols to form a hydrated silica gel,

Stage 4 Formation of an amorphous calcium phosphate layer,

Stage 5 Crystallization of a hydroxycarbonate apatite layer.

Stage 1 is usually controlled by diffusion and exhibits a $t^{-1/2}$ dependence, while Stage 2 is usually controlled by interfacial reaction and exhibits a $t^{1.0}$ dependence. The time for the onset of reaction stages for Bioglass® depends on the composition of the glass (Hench, 1991).

The literature base for the reactions is extensive, and numerous glass compositions have been studied. These five reaction stages do not depend on the presence of tissues. They occur in distilled water, tris-buffered solution or simulated body fluids. Bioglass® 45S5 has been used as a baseline for most surface studies, largely because it is a single phase glass and has only four constituent compounds. Results obtained from various investigations confirmed the above reactions (Kokubo, 1990; Andersson, 1990; Fowler, 1974).

The rate of tissue bonding appears to depend on the rate of HCA formation. However, bonding to tissues requires an additional series of reactions, which are less well defined. Biological understanding of these processes is needed and requires further investigation. At present, the following sequence of events appears to be associated with the formation of a bond with tissues is:

Stage 6 Adsorption of biological moieties in the SiO_2 – HCA layer,

Stage 7 Action of macrophages,

Stage 8 Attachment of stem cells,

Stage 9 Differentiation of stem cells,

Stage 10 Generation of matrix,

Stage 11 Mineralization of matrix.

The strong bond formed between the Bioglass® implant and bone is clearly demonstrated when during impact torsional loading fracture occurred in the implant, rather than at the implant – bone interface (Hench, 1991). Using conventional push – out tests, bond strength values of 15 – 25 MPa were obtained for a number of bioactive glasses (Hench and Wilson, 1993).

Rate of Bonding

The rate of development of the interfacial bond between implant and bone can be referred to as the level of bioactivity. Hench introduced an index of bioactivity (Hench, 1991), which is given by

$$I_B = 100/t_{0.5bb} \quad (1.15)$$

where $t_{0.5bb}$ is the time for more than 50% of the implant surface to be bonded to bone. In the ternary diagram for the compositional dependence of the bioactivity, iso-I_B contours are shown (Fig. 1.9). The closer to the bioactivity boundary a glass is, the slower is the rate of bonding. Bioglass® implants with the highest I_B values formed a mechanically strong bond to soft connective tissues as well as bone (Wilson and Nolletti, 1990). Bioactive implants with intermediate I_B values do not develop a stable soft tissue bond; instead, the fibrous interface progressively mineralizes to form bone. Consequently, there is a critical iso-I_B boundary beyond which bioactivity is restricted to stable bone bonding.

The thickness of the bonding zone is roughly proportional to the I_B value, and the failure strength of a tissue – implant bond appears to be inversely proportional to the thickness of the zone. The high degree of bioactivity of Bioglass® can stimulate the repair process and induce osteogenesis.

1.3.1.5 Clinical Applications

Applications of Bioglass® require several different forms of the material: (a) solid shapes, (b) particulates of various size ranges, (c) particulates combined with autologous bone particles; and (d) particulates delivered via an injectable system. Bioglass® does not break up when drilled with standard surgical drilling equipment. The clinical use of Bioglass® 45S5 in periodontal repair, maxillofacial reconstruction, and other areas has been reported (Wilson and Merwin, 1988; Hench, 1991; Wilson and Low, 1992).

The use of Bioglass® 45S5 implants in middle ear surgery to replace ossicles damaged by chronic infection is also successful, and results are encouraging (Reck et al., 1988; Merwin, 1990; Wilson et al., 1995). Bioglass® middle ear implants are shown in Fig. 1. 10. Problems encountered with plastic or metal implants can be solved by the special properties of Bioglass® 45S5 which provides implant immobilization and prevents implant extrusion. A bond with bone can be achieved with the staple of the implant. The soft tissue bond between the implant and the tympanic membrane eliminates movement at the interface and thus prevents implant extrusion. The stem of the implant may be shortened and shaped to fit the individual patient.

Bioglass® 45S5 implants have been used successfully for maintenance of the alveolar ridge for denture wearers with nearly a 90% retention rate (Hench and Wilson, 1993). The endosseous ridge maintenance implant (ERMI), a cone-shaped Bioglass® device is used to fill the defect in the jaw when a tooth is removed. Removal of one or two teeth produces changes in the jaw bone which are followed by gradual bone loss, so that the normal shape of the bone which supports healthy teeth changes to a narrow knife-edge ridge with reduced height which cannot comfortably support dentures. Without a means of preventing this bone loss, denture wearers are often destined to suffer increasing discomfort from ill-fitted dentures and in many cases may eventually become unable to wear dentures at all. Bioglass® ERMI was shown to act under a time-dependent kinetic modification of its surface after placement. Within 1 hour after implantation, formation of a chemical bond of Bioglass® with bone appeared to have occurred. Bioglass® devices have proved to be more successful than other materials which were tried (Stanley et al., 1987).

Figure 1.10 Bioglass® middle ear devices

1.3.2 Cerabone®

1.3.2.1 Composition and Structure

Since the discovery of Bioglass® by Hench in the early 1970s, various glasses and glass-ceramics have also been found to bond to bone, and some of them are now in clinical use. These glasses and glasses have different compositions, constituent phases, and hence different properties. One of the prominent bioactive glass-ceramics is Cerabone® which was developed by Kokubo and co-workers in the early 1980s (Kokubo et al., 1982). Cerabone® is a glass-ceramic which contains apatite and β-wollastonite (CaO · SiO).

Cerabone® can be prepared from the parent glass in the pseudoternary system $3CaO \cdot P_2O_5 - CaO \cdot SiO_2 - MgO \cdot CaO \cdot 2SiO_2$. Both the apatite and wollastonite are homogeneously distributed in a glass matrix, taking the shape of rice grains 50 – 100 nm in size. According to powder X-ray diffraction, the contents of apatite, wollastonite, and residual glass phase were 38 wt%, 34 wt%, and 28 wt%, respectively, and the composition of the residual glass phase was estimated to be MgO 16.6 wt%, CaO 24.2 wt%, and SiO_2 59.2 wt% (Kokubo et al., 1992). The glass-ceramic is named A-W glass-ceramic after the crystalline phases in it and is called Cerabone® A-W commercially.

1.3.2.2 Processing

Glass-ceramics are made by controlled crystallization of glasses. Several apatite-containing glasses were attempted initially. For A-W glass-ceramic, a parent glass in the pseudoternary system $3CaO \cdot P_2O_5 - CaO \cdot SiO_2 - MgO \cdot CaO \cdot 2SiO_2$ was prepared by the conventional melt-quenching method. The powder compact of a glass of the nominal composition MgO 4.6 wt%, CaO 44.7 wt%, SiO_2 34.0 wt%, P_2O_5 16.2 wt%, CaF_2 0.5 wt% was heated to 1050℃ at a rate of 1℃/min, held at 1050℃ for 4 h, and allowed to cool in the furnace (Kokubo et al., 1986). As a result, the glass powder was fully densified at about 830℃, and then oxyfluoroapatite and β-wollastonite were precipitated successfully at 870℃ and 900℃, respectively, to give a crack- and pore-free, dense and homogeneous glass-ceramic. β-Wollastonite consisting of a silicate chain structure was chosen as the reinforcing phase. Important factors in forming glass-ceramics are the nucleation and the growth of crystals of small, uniform sizes.

1.3.2.3 Properties

A-W glass-ceramic can be easily machined into various shapes, even into screws, by using diamond tools. Commercially available Cerabone® is shaped into artificial vertebrae, intervertebral spacers, spinous process spacers and iliac spacers. The physical properties of A-W glass ceramic are shown in Table 1.8. The bending strength of the glass-ceramic is about 200 MPa in the atmospheric environment, which is almost twice that of dense sintered hydroxyapatite (115 MPa) in clinical use and even higher than that (160 MPa) of human cortical bone. The parent glass precipitating only the apatite has a bending strength of 72 MPa. It is evident that the higher bending strength of the glass-ceramic is due to the precipitation of wollastonite. A-W glass-ceramic has a fracture toughness value of 2.0 $MPa \cdot m^{1/2}$ while its parent glass only has 0.80 $MPa \cdot m^{1/2}$. The parent glass shows a fairly smooth fracture surface, whereas A-W glass-ceramic has a roughened fracture surface. This indicates that the wollastonite phase effectively prevents straight propagation of cracks, deflecting them or causing them to branch out. It is notable that even though it is not in the fibrous form, the reinforcing wollastonite phase shows a toughening effect. A-W glass-ceramic has the highest mechanical strength among all bioactive glasses and glass-ceramics developed (Kokubo et al., 1985) while

Table 1.8 Physical properties of A-W glass-ceramic

Density /g/cm^3	3.07
Young's modulus /GPa	118
Bending strength /MPa	215
Compressive strength /MPa	1080
Vickers hardness (VHN)	680
Fracture toughness /$MPa \cdot m^{1/2}$	2.00

exhibiting a fairly high degree of bioactivity (Nakamura et al., 1985).

Under the loading conditions in an aqueous environment, A-W glass-ceramic showed a decrease in strength, which is due to slow crack growth caused by stress corrosion. The magnitude of the decrease, however, was much lower than that of its parent glass (Kokubo et al., 1987). According to an estimats, in a simulated body fluid with an ion concentration nearly equal to that of human blood plasma at 36.5℃, it is expected that A-W glass-ceramic could withstand continuous loading with a bending stress of 65 MPa for over 10 years, whereas its parent glass, a glass-ceramic containing only apatite and a sintered dense hydroxyapatite all might be broken within one minute (Kokubo et al., 1987). Animal experiments have shown that A-W glass-ceramic implants can maintain their high strength for long periods *in vivo* (Kitsugi et al., 1987a).

1.3.2.4 Surface Apatite Formation and Bone-Bonding Mechanism

Electron probe X-ray microanalysis of the interface of A-W glass-ceramic with rabbit bone showed that a thin layer, about 10 μm thick and rich in Ca and P, was present at the interface (Kitsugi et al., 1986, 1987b). This Ca- and P-rich layer was composed almost entirely of crystalline apatite, as shown by X-ray microdiffraction (Kokubo et al., 1990). The glass-ceramic was continuously connected to the living bone through this apatite layer without a distinct boundary between them (Neo et al., 1992). It appears that the essential condition for glasses and glass-ceramics to bond to living bone is the formation of a certain type of biologically active apatite layer on their surfaces.

The same type of apatite layer can be formed on the surfaces of bioactive glasses and glass-ceramics even in an acellular simulated body fluid with ion concentrations nearly equal to those of human blood plasma of pH 7.25 at 36.5℃ (Table 1.9) (Kokubo, 1993). The surface structure of $CaO \cdot SiO_2$-based glasses provides favorable sites for apatite nucleation. These glasses form a silica hydrogel layer prior to the formation of the apatite layer (Ohura et al., 1991; Ohtsuki et al., 1992). It is probable that the hydrated silica induces apatite nucleation. The mechanism of apatite formation on the surfaces of $CaO \cdot SiO_2$-based glasses and glass-ceramics, including A-W glass-ceramic, in the body can be postulated and summarised as follows. The calcium ions dissolved from the glasses and glass-ceramic increases the ion activity product of the apatite in the surrounding body fluid, and the hydrated silica on the surface of the glasses and glass-ceramics provides favorable sites for apatite nucleation. Consequently, the apatite nuclei are rapidly formed on their surfaces. Once the apatite nuclei are formed, they spontaneously grow by consuming calcium and phosphate ions from the surrounding body fluid (Fig. 1.11) (Kokubo, 1993). Although the presence of the silica gel layer could not be detected on the A-W glass-ceramic surface even under high resolution TEM, the dissolution of an appreciable amount of the silicate ion from the glass-ceramic into the fluid indicated the formation of a large number of silanol groups on the glass-ceramic surface in the body.

Table 1.9 Ion concentrations in simulated body fluid (SBF) and human blood plasma

	Ion Concentration /mM							
	Na^+	K^+	Ca^{2+}	Mg^{2+}	Cl^-	HCO_3^-	HPO_4^{2-}	SO_4^{2-}
SBF	142.0	5.0	2.5	1.5	147.8	4.2	1.0	0.5
Human plasma	142.0	5.0	2.5	1.5	103.0	27.0	1.0	0.5

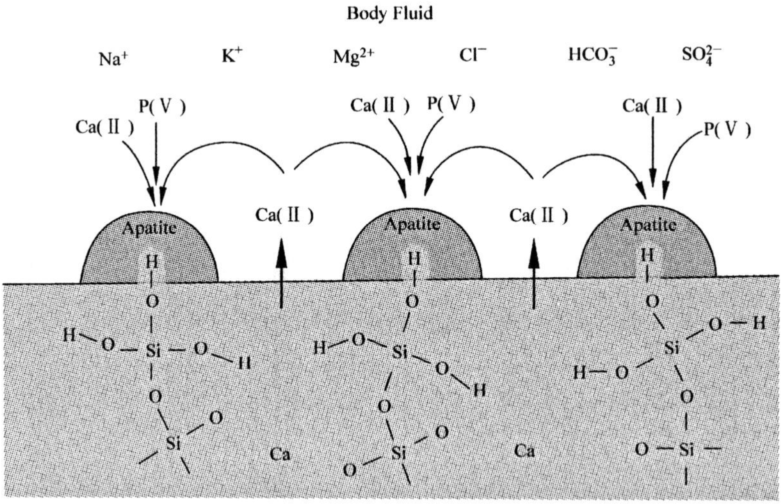

Figure 1.11 Schematic representation of the mechanism of apatite formation on the surfaces of CaO · SiO_2-based glasses and glass-ceramics in the body

A thin layer rich in Ca and P, but poor in Mg and Si, was always observed at the interface between the implanted A-W glass-ceramic and bone. Its thickness varied from 0.5 μm to 100 μm depending on the type of animal, the type of bone, and the period of implantation. The compositional and structural characteristics of crystallites in this layer indicate that they are similar to bone apatite. Therefore, once an apatite layer is formed *in vivo* after implantation, it can be expected that osteoblasts proliferate on the surface of the apatite layer in preference to fibroblasts and consequently the surrounding bone comes into direct contact with the glass-ceramic without the intervention of the fibrous tissue. When this occurs, a strong chemical bond is formed between the surface apatite and the bone apatite (Kokubo, 1991). A-W glass-ceramic is observed to be tightly bonded to living bone after implantation. During detaching tests, fracture under tensile load usually did not occur at the implant – bone interface. It took place within the bone (Nakamura et al., 1985).

1.3.2.5 Clinical Applications

Since 1983, A-W glass ceramic has been used in spine and hip surgery of patients with extensive lesions or bone defects. Satisfactory results have been reported (Yamamuro, 1993). A-W glass-ceramic vertebral prostheses have different sizes, so that the surgeon is able to choose the appropriate one in the operating theater. A vertebral prosthesis in combination with some autogenous bone graft can be securely implanted in the bone defect. One such implant reportedly survived up to 8 years without causing recurrence of the tumor in the operated site, although the patient needed replacement of the prosthesis due to other reasons. In cases of aseptic loosening of hip prostheses combined with large bone loss, A-W glass-ceramic granules can be used as a bone defect filler. In such cases, the glass-ceramic granules are combined with autogenous cancellous bone and fibrin glue to fill the defects. The glass-ceramic-fibrin mixture showed significantly better osteoconduction and acceleration of the bone repairing process than glass-ceramic granules alone. Loosening of the stem of hip prostheses has been revised by the use of a long stem in combination with A-W glass-ceramic granules mixed with autogenous bone and fibrin glue. The amount of A-W glass-ceramic granules used in the operation ranged from 10 g to 75 g.

In surgery of the spine, skeletal tumors, and skeletal trauma, large autogenous bone grafts are often taken from the iliac crest. To prevent various clinical complications and fill bone defects, iliac crest prostheses of various sizes are fabricated from A-W glass-ceramic. Clinical results 1 to 2 years after operation were excellent with 97% of the patients. New bone formation around the iliac crest prostheses progressed steadily, and 1 year following the operation, good new bone formation was observed. In half of the clinical cases, there was no radiological clear zone on either side of the prosthesis (Yamamuro, 1990).

1.3.3 Other Bioactive Glasses and Glass-Ceramics

1.3.3.1 BioGran®

BioGran® is a bone graft material made of bioactive glass granules (300 – 355 μm) which contain calcium, phosphorous, silicon, and sodium. This material is hydrophilic and slightly hemostatic. When mixed with sterile saline or the patient's blood, a cohesive mass is formed that can be shaped to fill the defect. It has been shown extensively that bone formation and growth occur within each glass granule. This osteogenesis, guided by bioactive glass particles, occurs at multiple sites, rapidly filling the osseous defect with new bone that continuously remodels in the normal physiologica manner (Schepers et al.,

1991). The controlled bioactivity permits material resorption and bone formation to occur simultaneously. Eventually, the graft is resorbed, with animal studies showing that the material may be replaced by new bone (Schepers et al., 1991; Schepers and Pinruethal, 1993; Schepers and Ducheyne, 1993).

BioGran® glass particles were put in bone defects in beagle dogs (Shepers and Ducheyne, 1997). It was shown that repair of the defects was enhanced by the osteoconductive properties of these particles. In addition, a protective pouch, in which bone formation was separated from the external bone tissue, was formed. BioGran® was used for subantral augmentation after sinus elevation (Furusawa and Mizunuma, 1997). Histological analysis revealed new bone formation in all cases. The biomechanical properties of the regenerated bone and those of native bone tissue were similar.

1.3.3.2 Ceravital®

After Hench's invention of Bioglass®, a family of glass and glass-ceramic was designed by Bromer et al. for bone and tooth replacement (Bromer et al., 1973). These glasses and ceramics, called "Ceravital", contain mainly SiO_2, $Ca(PO_2)_2$, and CaO with small amounts of Na_2O, MgO, K_2O, and other oxides. Even though the original idea was to use these new materials in load-bearing conditions for the replacement of bone, their mechanical properties were not found compatible for such applications. It was also shown that the surface reactivity was operative in areas of the implant surface where soft tissue was interfacing, which led to dissolution of the material and activation of mononuclear and multinuclear resorbing cells. This caused concerns about the long-term stability of the material. Therefore, some other glass compositions were used in order to decrease the solubility of Ceravital® glasses and glass-ceramics (Gross and Strunz, 1980, 1985 Gross et al., 1981). *In vitro* experiments showed that this approach was very successful. The solubility of the material could be adjusted to various degrees by the addition of metal oxides to the melt. However, *in vivo* investigations also indicated that materials containing metal oxides were not conducive to mineralization of the bone matrix adjacent to the implant interface. In order to understand the mechanisms of mineralization and the inhibition of mineralization on bone-bonding and nonbonding Ceravital® glasses and glass-ceramics, the mineralization process was investigated and the role of matrix vesicles in the promotion of mineral deposition analyzed (Muller-Mai et al., 1990, 1991). Even though Ceravital® of a number of compositions was investigated *in vitro* for its behavior regarding solubility in water and in physiological saline solution, only a selection of these compositions was investigated *in vivo*. *In vivo* tests used different animal models including mouse, rat, and rabbit. It was shown that for different species there was a very close resemblance of the cellular reaction and the behavior of the cellular matrix

(Gross et al., 1993). The cellular reaction was different for glasses and glass-ceramics, depending on the amount of metal oxides being released. Several approaches have been investigated for improving the biomechanical properties of Ceravital®. In one series of experiments, metal surfaces were coated with enamel onto which particles of bone-bonding glass-ceramic were planted. The bone-bonding glass-ceramic particles were found to be covered and anchored with bone. However, the surfaces between these particles were enamel which contained various metal oxides, which caused inhibition of mineralization of bone matrix (Zeiler et al., 1986). In another series of experiments, composites containing glass-ceramic particles and metal were produced. The metal was titanium and composites with various percentages of glass-ceramic particles were tested *in vivo*. These tests gave some promising results (Muller-Mai et al., 1989).

The only clinical application of Ceravital® to date is the replacement of the ossicular chain in the middle ear where the loads are minimal and the mechanical properties of the material are adequate (Reck, 1984).

1.3.3.3 BIOVERIT

From different base glasses, BIOVERIT glass-ceramics have been developed as machinable and bioactive biomaterials (Vogel and Holand, 1987; Holand et al., 1991). BIOVERIT Ⅰ is a mica-apatite glass-ceramic with a chemical composition from the SiO_2 – (Al_2O_3) – MgO – Na_2O – K_2O – F – CaO – P_2O_5 base glass system. BIOVERIT Ⅱ glass-ceramic contains mica as the main crystal phase and secondary crystals (e.g. cordierite). BIOVERIT Ⅲ is SiO_2 – free and bioactive, and contains apatite, $AlPO_4$, and complex phosphate structures. All these glass-ceramics have good chemical properties (hydrolytic stability). The adequate mechanical properties of BIOVERIT Ⅰ and Ⅱ allow them to be used as biomaterials for bone substitution. BIOVERIT Ⅲ can be used in composites with certain metals. More than 850 implants made of BIOVERIT Ⅰ and Ⅱ have been successfully used in clinical cases in the fields of orthopedic surgery, head and neck surgery, and stomatology (Holand and Vogel, 1993).

1.3.3.4 Apoceram

Another glass-ceramic, Apoceram, was developed as a bioactive, cast material (Carpenter et al., 1986). This material is based on the Na_2O – CaO – Al_2O_3 – SiO_2 system with P_2O_5 as a nucleating agent and consists of interlocking spherulites of apatite and wollastonite crystals. Fabrication of the material utilizes the powder route and hot pressing, at a temperature at which both sintering and crystallization take place. However, the toughness of this material is not adequate for high-load implant applications.

1.4 Bioactive Coatings

1.4.1 Plasma Sprayed Hydroxyapatite Coatings

1.4.1.1 Thermal Spray Techniques

Thermal spraying processes utilize a high-energy heat source to melt and to accelerate fine particles onto a prepared surface. Upon impact these particles cool down and resolidify instantly by heat transfer to the underlying substrate and therefore form, by accumulation, a coating consisting of lamellae. Thermal spray processing has been used for many years to deposit layered coatings for various purposes such as wear resistance, thermal barrier, biocompatibility, etc. The major advantages of the thermal spray processes are their ability to apply a very wide variety of compositions including most metallic, ceramic, cermet, and some polymeric materials and the ability to do so without significantly heating the substrate surface to be coated (Tucker, 1994). The main thermal spraying techniques include flame spraying, plasma spraying, high velocity oxygen fuel (HVOF) spraying, vacuum plasma spraying, arc metallization, and detonation gun spraying. Among them, flame spraying, HVOF spraying, and plasma spraying are used in the production of calcium phosphate coatings.

Combustion Flame Spraying
In flame spraying, the flame is generated by the controlled ignition of a combustible fuel gas and oxygen (de Groot et al., 1990). The flame melts the powders which are injected by compressed air. This form the spray streams that successively propel the molten powders toward the substrate. The bond strength can reach 60 MPa for flame-sprayed Ni – Al coatings, or even 70 MPa for self-fluxing alloys. Typical values are in the range of 15 MPa for ceramics coatings to 30 MPa for other materials. Porosity in the coatings is in the range of 10% – 20%.

High Velocity Oxy-Fuel (HVOF) Spraying
In the HVOF process, the fuel (e.g., hydrogen, acetylene) is burned with oxygen at high pressure and generates a high velocity exhaust jet. The ceramic powder is injected axially into the jet as a suspension in the carrier gas. The temperature of oxygen – acetylene reaches a maximum at 3440.0 K if mixed at the ratio of 1.5 to 1.0 (by volume). The gas velocities can reach 2000 m/s. This makes it capable of accelerating the injected particles much more rapidly than a normal flame and can provide a potentially distinct advantage for the development of highly adhesive coatings. Compared with other thermally sprayed coatings, HVOF coating has several advantages: high density, excellent hardness, wear resistance, good corrosion resistance, and high bond strength.

The bond strength of HVOF sprayed carbides can be as high as 90 MPa. The porosity of HVOF sprayed coatings is lower than 1%.

Plasma Spraying

The process of spraying with the use of plasma was patented in the early 1960s. The technical creation and utilization of plasma as a high-temperature source is realized in the plasma torch, or the plasma gun (Pawlowski, 1995). The plasma gun consists of a cone-shaped thoriated tungsten cathode and a cylindrical copper anode. Gases flow through the annular space between the two electrodes, and an arc is initiated by a high-frequency discharge. The stream of gas which flows between the two electrodes stretches the arc, so that in its course from one electrode to the other, the arc loops out of the nozzle of the gun as a plasma flame. The plasma temperatures in spraying guns are normally in the range of 10,000 – 15,000℃ but can reach 30,000℃ in certain devices. This implies that, in principle, almost any material including refractory metals or oxides can be melted and deposited by plasma spraying.

In plasma spraying, feedstock materials (in the form of powder, rod, or wire) are introduced into the plasma flame to be melted. The melted materials arrive on the surface of target after having been sufficiently heated and accelerated by the plasma jet. When the droplets impact the target, they flatten, spread out on the surfaces, and form a deposit through successive impingement (Fig. 1.12) (Lacefield, 1993). At present, gases such as Ar, He, H_2, and N_2 are mainly used as the arc gases (plasma-forming gases). The velocity and temperature of the powder particles are directly related to gas type. Typical powder velocities for the conventional guns are in the range of 120 – 350 m/s, while for certain high velocity guns, values of 400 – 550 m/s can be obtained.

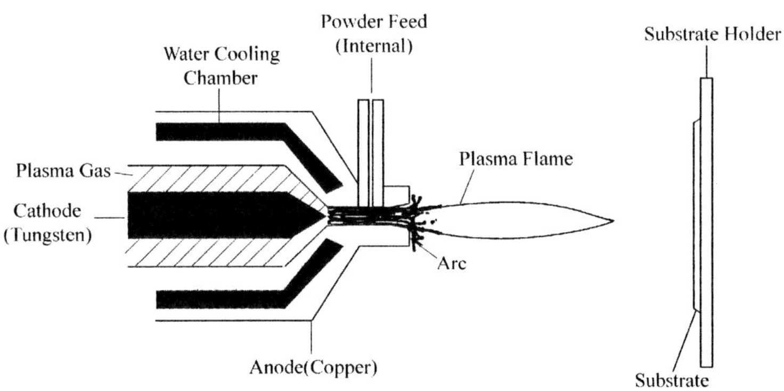

Figure 1. 12 Schematic diagram showing the plasma spraying process

Most powders used for plasma spraying have particles between 10 μm and 90 μm in diameter. Powders with this narrow size distribution are preferred to achieve uniform heating and acceleration of a single component powder. Small particles can be evaporated in the flame, and their trajectory may be so dominated by the momentum of the plasma gas that they follow the gas stream around the substrate rather than impacting it. Large particles, on the other hand, may not be melted or may fall from the flame or rebound from the target. A constant powder feed rate is a prime condition for achieving uniform coating thickness and quality. The powders to be injected into the plasma must possess good flow properties, which is again associated with their morphology and size. How effectively a given plasma flame heats and accelerates powder particles depends on the coating material itself, i. e. its composition, density, heat capacity, conductivity, size, and shape of the particles. Due to the large number of parameters involved in the plasma spray process and the complex relationships that exist among these parameters, precise control and optimization of the deposition process is a very tedious and expensive operation. The kinetic and thermal energy that particles gain and retain in the plasma jet determine the coating structure and the coating properties.

Plasma spray deposition involves two separate processes (Table 1. 10): plasma – particle interaction and particle – substrate interaction. The solidification process has been studied extensively using scanning and transmission electron microscopy, and it was suggested that as a splat solidifies, heat is lost to the substrate beginning at the center of the splat (Herman, 1988). A solid core forms, and the remaining melt spills off and hardens into a raised rim. The particles impacting the substrate can be solid (not molten), softened, semiliquid, or liquid. The types of splats formed strongly determine the final coating structure and properties.

Table 1.10 Mechanism in the plasma spraying process

Particle-Plasma Interaction	Particle-Substrate Interaction
Injection	Impact
Acceleration	Spreading
Melting	Rapid solidification
Superheating	Bonding

The plasma sprayed coating is formed by the buildup of successive layers of particle droplets flattened on impact (Fig. 1. 13) (Herman, 1988). Molten particles spread out and interlock with one another. Voids resulting from shrinkage or from trapped air are present. Some particles have become oxidized during their trajectory, while the others may not have melted and simply get embedded in the deposit. Cooling rates upon impact are estimated to be of the order of $10^6 - 10^8$ K · s^{-1}. The process of coating is based on the buildup of rapidly solidification splats and the development of a highly defective microstructure. Another result of rapid solidification is the formation of a

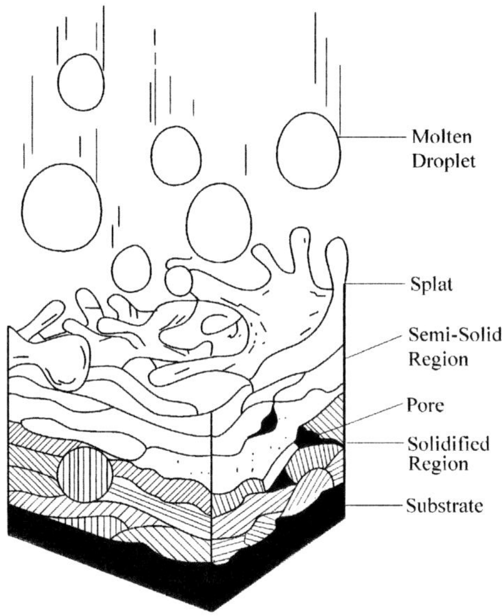

Figure 1. 13 Buildup of a defective coating by the deposition of molten droplets in the plasma spraying process

glass or an amorphous structure.

1.4.1.2 Plasma Sprayed Calcium Phosphate Coatings

Investigations into calcium phosphate ceramic coatings on metallic implants started with the observation that HA in the pores of a metal implant with a porous coating would significantly affect the rate and vitality of bone ingrowth into the pores (Ducheyne et al., 1980). In 1987, de Groot et al. reported the plasma sprayed HA coating on Ti-6Al-4V cylindrical rods (de Groot et al., 1987). Since then, many experimental and clinical studies on plasma sprayed HA coatings have been conducted.

Characteristics of Plasma Sprayed Calcium Phosphate Coatings

In the biomedical field, coatings have been used to modify the surface of implants and, in some cases, to create an entirely new surface, which gives the implant properties that are quite different from the uncoated device. One reason for the use of HA or a similar calcium phosphate surface is to cause early stabilization of the implant in surrounding bone, as in the case of a dental implant where the healing time is reduced. Another reason to use an HA coating is to extend the functional life of the prosthesis, as in the case of a cementless hip prosthesis stabilized by the HA coating in the surrounding femur.

Currently, the most extensively used method for depositing a HA coating is plasma spraying. However, this deposition process involves high temperature which has the possibility of dehydroxylation and decomposition of HA in the coating. The XRD pattern of a plasma sprayed HA coating shows the presence of other phases in addition to crystalline HA. These new phases include α-tricalcium phosphate (α-TCP), β-tricalcium phosphate (β-TCP), tetracalcium phosphate (TTCP), and calcium oxide (CaO). The reduction in peak intensity and peak broadening of HA peaks suggest the formation of amorphous calcium phosphate (Zyman et al., 1993). The formation of these phases is a result of extreme temperatures, rapid cooling, and highly reactive atmospheres that favor nonequilibrium or metastable structures. It was also reported that the crystalline size of HA was considerably smaller than that in conventionally sintered dense HA (Koch et al., 1990). Nevertheless, it is well known that plasma sprayed HA coatings have the ability to induce new bone ingrowth and subsequently improve fixation stability. Clinical evidence has indicated that the presence of HA can promote osteointegration between bone and implant devices (Rivero and Fox, 1988). Although several reports indicated that HA coating might degrade and fail either at the interface or on the surface, work involved in the development of a satisfactory HA coating has received worldwide attention.

Considering the sensitive nature of the biological environment, the HA coating of a complex microstructure produced by the plasma spray process is certain to induce a range of tissue responses significant in coating performance. The solubilities of HA, TCP, and amorphous calcium phosphate in body fluids are known to be different. Crystalline HA is considered nonresorbable, whereas TCP is resorbable and amorphous calcium phosphate is highly resorbable. This suggests that the formation of resorbable TCP and amorphous calcium phosphate after plasma spraying of HA greatly alters the biological response. Studies carried out *in vitro* and *in vivo* also revealed that the biodegradation of HA coatings depended on their crystallinity which refers to the percentage of crystalline phase in the coating (Mendelon et al., 1994). Some researchers believed that the intended function of a thin protective bioactive coating could become severely impaired if amorphous calcium phosphate were formed instead of crystalline HA (Wolke et al., 1990). On the other hand, it was found that the amount of newly formed bone on the α-TCP surface was more than that on the HA surface. Therefore, if coating longevity is desired, then the crystalline HA coating is preferable, and if only the osteoconductive property of the calcium phosphate coating is desired for initial fixation, the resorbable calcium phosphate coating is advantageous. The quantity of amorphous and resorbable phases in HA coatings is expected to change, depending on the spraying conditions. Therefore, to meet different medical applications, it is desirable to tailor the crystallinity of plasma sprayed HA coating by carefully choosing the feedstock powder, varying the plasma spraying atmosphere, and controlling the plasma spraying parameters so as to adjust particle melting and flow conditions. It was suggested that the stoichiometric Ca/P molar ratio and crystallinity of starting HA powders play a key role in determining the crystallographic structure of the resulting HA coating (de Groot, 1980). Particle morphology, particle size, and size distribution are also important factors.

Another important factor when considering the desirable properties of a coating for implants is that the coating should be strongly bonded to the metal substrate so as to maintain implant integrity as well to facilitate proper transmission of the load from the implant to the surrounding bone. A HA coating which separates from the implant *in vivo* would provide no advantage over an uncoated implant and may be less desirable than no coating at all. In a worst case condition, a weakly bonded HA coating may separate from the implant, and fragments of the coating would be in close proximity to the bare metal surface. Any movement of the loose ceramic coating fragments on the substrate surface could result in disruption of the passive oxide layer on metal surfaces, such as TiO_2 on titanium. Although re-formation of the oxide layer occurs quite rapidly, a momentary increase in metallic ion release will take place. Furthermore, the oxide which forms under *in vivo* conditions at the time of re-formation may not be as passive or protective as the original oxide layer. Therefore, good adhesion of a HA coating to a metal substrate is of prime importance. This prevents breakdown of tissue attachment to the implant and enables it to be used for long periods of time.

HA has been used to coat Ti-6Al-4V hip prostheses. The coating is applied at the stem because this part interfaces with the osseous tissue if the coating is absent. Ti-6Al-4V as a substrate is more favorable compared to other alloys such as 316 L stainless steel and Co – Cr alloys. It has an elastic modulus and density that are more compatible with bone. Compatibility between elastic moduli of bone and that of the Ti alloy is important in order to avoid a difference in deformation between bone and the implant when stress is applied. Mechanical incompatibility of this nature will be transferred to the interface in the form of deformation stresses and strains, while the interface is a weak area. Ti-6Al-4V also has a very low modulus mismatch with that of HA. This minimizes the deformation stresses at coating – substrate when the implant is loaded (Ravaglioli and Kraejewski, 1992).

Postspray Treatment of Calcium Phosphate Coatings

Three types of postspray treatment are often used for improving the properties of plasma sprayed calcium phosphate coatings: hot isostatic pressing (HIPing), laser surface treatment, and heat treatment. The post-coating heat treatment is normally accomplished under vacuum or an inert environment to produce thin interdiffusion layers between the coating and substrate. Apart from providing enhanced bonding, the treatment also promotes recrystalliation of the original lamellar structure, leading to homogenization of the coating composition and some reduction in residual porosity. It was reported that heat treatment of amorphous coatings at 800℃ for two hours caused a fourfold rise in XRD peak height, suggesting an increase in the crystallinity of HA (Berndt and Gross, 1992). After heat treatment, tricalcium phosphate and tetracalcium phosphate phases were no longer detected by XRD analysis in heat treated coatings.

HIPing involves the simultaneous application of isostatic pressure and high temperature to a workpiece, leading to the consolidation of the workpiece. It was shown that by post coating HIPing, porosity levels in plasma sprayed coatings could be effectively reduced to 4% (Shankar et al., 1981). The major disadvantage of HIPing is its high cost.

Laser treatment may enable the elimination of porosity, enhance coating strength and chemical homogeneity, and improve coating adhesion. A pulsed YAG laser was used to treat the surface of plasma sprayed HA coatings (Khor and Cheang, 1994). The results showed that a pulsed laser can effectively melt the surfaces of HA coatings, creating a surface with a morphology different from that of the original coating. The laser treated surface was characterized by a smoother surface, spherically shaped pores, crack networks, and fine grains. The amount of cracks and pores increased as the pulse energy level increased.

1.4.1.3 Mechanism of Adhesion in Thermally Sprayed Coating

The use of plasma sprayed HA implants for medical applications is strongly influenced by the long-term stability of the coating system. The coating – substrate adhesion and interlamellar cohesion are of great concern since they ultimately govern the effectiveness of the sprayed deposit as a bioactive coating. The operative interparticle and coating – substrate adhesion mechanisms of plasma sprayed coatings have been classified into three major categories.

Mechanical Bonding or Interlocking

Many lamellae on the roughened surface are, to some degree, attached to the substrate by the force resulting from the shrinkage of the liquid wrapping around the surface irregularities. The lamellae do not contact the substrate at all of their bottom surfaces. The areas of contact are called "welding points" or "active zones" and correspond to 20% – 30% of the total area of the lamella. The adhesion of the coating improves if the contact area becomes greater. Inside the contact areas between the lamellae and the substrate, adhesion could result from physical interaction, metallurgical interaction, and other interactions.

Physical Interaction

The physical interaction between the atoms of the lamella and those of the substrate results from the action of van der Waals forces and can occur only if there is close and constant contact. The conditions prerequisite to the physical interaction between the contacting surfaces are as follows:

(1) Surfaces are clean;

(2) Surfaces are in a state of higher energy (e.g. by plastic deformation);

(3) The contact is close (this occurs more readily if the lamella is liquid).

Metallurgical Interaction

There are two possible mechanisms of metallurgical interaction: diffusion and chemical reaction between lamella and substrate. The high contact temperature increases the probability of diffusion. The formation of a chemical compound between the sprayed particles and the substrate also gives very strong metallurgical bonding. The bonding between coating and substrate or between individual splats is generally the result of two or more bonding mechanisms operating simultaneously. However, the degree to which a particular mechanism is operative strongly depends on the materials involved and the process variables employed.

1.4.1.4 Testing Method for the Adhesion of Thermally Sprayed Coatings

The evaluation of the interfacial strength between coatings and substrates is of great importance to nearly all branches of surface engineering. The American Society for Testing and Materials (ASTM) defines adhesion as "the state in which two surfaces are held together by interfacial forces which may consist of valence forces or interlocking forces or both" (ASTM D907-70). There is a distinction between basic adhesion, which is the maximum possible attainable value, and experimental or practical adhesion, which can be termed bond strength or adhesion strength (Mittal, 1976). The measured adhesion is much smaller than the basic adhesion owing to the effects of various factors which include internal stresses in the coating and the testing technique used.

Adhesion can be experimentally measured in two ways: (1) in terms of forces, defining the force of adhesion as the maximum force per unit area; and (2) in terms of work or energy, defining the work of adhesion as the work done in separating or detaching two materials from one an other. Methods for the measurement of coating – substrate adhesion are listed in Table 1.11 (Rickerby, 1988). Many other methods are used to evaluate adhesion qualitatively or quantitatively . Each method has disadvantages , and none can elucidate the

Table 1.11 Methods for determining coating adhesion

Qualitative Method	Quantitative Method
Mechanical method	Mechanical method
Cellophane tape test	Tensile bond test
Abrasion test	Shear bond test
Scratch test	Indentation test
	Scratch test
Nonmechanical method	Nonmechanical method
X-ray diffraction	Thermal method
	Ultracentrifugal test
	Nucleation test
	Laser spallation test
	Capacitance measurement

differences of in-service performance. For thermally sprayed coatings, testing methods are restricted. One difficulty with any mechanical property assessment of coatings involves attachment of a loading device to the coating without influencing the property that is being measured. No testing method covers all situations, and no adhesion test satisfies all requirements. Therefore, the best method is often the one that simulates practical stress conditions.

Several adhesion tests have been reviewed (Lin and Berndt, 1994). Three general classes of tests can be considered, each of which has its merits and problems. Several test methods, such as ASTM C633-79, detect the load or stress required to fail a standard joint. Most of these tests require a strong bonding agent to attach the coating to a loading fixture. Tensile adhesion tests suffer from epoxy penetration and alignment problems. Another feature of both the tensile and the shear bond tests is that the fracture mode is ambiguous, because mixed-mode failures often occur. Mixed-mode failure, where a single fracture surface exhibits both adhesive and cohesive components, is a major practical obstacle in coating development. The inability to perform simple tests that reflect distinct fracture morphologies makes the physical interpretation of current adhesion tests difficult. The main advantage of these tests is that they are simple to perform and the calculation of the strength is straightforward. They are therefore the most popular methods used.

The adhesion of thermally sprayed coatings is not only an interfacial problem of each lamella within a coating but also involves the integrity of the interface between the substrate and coating, residual stresses, crack population, and pore size and distribution. Fracture mechanics considers the energy required to initiate or propagate cracks and evaluates the adhesion of the coating system in terms of fracture toughness. Methods of measurements include the bend test (three or four point), single-edge notch test, compact tension test, etc. These tests usually attempt to detect the load at which a crack begins to propagate and relate that macroscopic load to the stress intensity at the crack tip. This requires knowledge of the elastic behavior of the entire system and usually makes assumptions about the elastic properties of the coating.

The third class of measurement techniques, perhaps the most desirable, are nondestructive techniques. By measuring adhesion directly on the engineered component, all of the considerations of how well the sample reflects reality are not necessary. The application of these techniques to thermally sprayed coatings is still relatively unexplored but several researchers have developed ultrasound techniques which successfully evaluated adhesion (Suga et al., 1992). These techniques suffer from the disadvantage that the measured parameters (usually wave intensity) are not useful in making quantitative predictions about service failure. A calibration procedure must be used to compare values obtained with measured ultrasound intensity. These techniques are, however, useful as quality monitoring methods, once a range of acceptable measurements is experimentally determined.

1.4.1.5 Factors Influencing Bond Strength of Thermally Sprayed Coatings

During tensile bond strength tests, if failure occurs at the coating – substrate interface, the parameters that influence the bond strength are correlated with

(1) the material used to thermal spray (especially the thermophysical properties of the material that determine the contact temperature, such as latent heat of fusion, specific heat, thermal conductivity and diffusivity), its size, and method of preparation;

(2) the processing technique and the selection of parameters;

(3) the substrate thermophysical properties (i. e. thermal diffusivity and conductivity) and the state of the surface (i. e. roughness and /or the way of roughening).

Substrate Surface Preparation

The shear strength of a bone – implant interface is significantly influenced by the surface roughness of the implant and is increased in a surface roughness-dependent manner. It was reported that HA-coated smooth implants had a weak implant – substrate interface and consequently a lower bond strength to bone than rough-surfaced implants (Nakashima, 1997). It is well known that the rougher the substrate, the rougher the HA coating surface. Therefore, to achieve good interfacial shear strength with bone, it is essential to roughen the substrate surface.

Substrate surface preparation is essential for the thermal spraying process. It not only involves cleaning the substrate surface, but also includes the important step of surface roughening. Normally, a sandblasting treatment, using cast iron grit, corundum sand (α-Al_2O_3), or silicon carbide (SiC), is performed on the substrate to remove any chemical or organic remains and any residual trace of oxides or mineral salts. Apart from sandblasting, other methods such as chemical etching and precoating can be used for roughening the substrate surface.

Splat Morphology

The parameters of the plasma spraying process and the properties of the coating materials and the substrate influence the splatting of the plasma sprayed particle on the substrate surface, which is the most fundamental process for coating formation. Two types of splat morphology can be formed: disk-shaped and splash-shaped. The splash splats are shown to affect adhesion adversely due to their rapid cooling rates, reduced fluid pressures, and interfacial voidage. The topography of the substrate can be used to suppress splashing by providing an additional means of dissipating the kinetic energy of the incoming droplets.

Plasma Spraying Parameters and Properties of Coating Materials and Substrate

The adhesion of sprayed coatings is also affected by the parameters of the plasma spraying process. Several parameters influence the spreading of a droplet. Among those, it has been shown that the impact velocity, the viscosity of the particle, the droplet temperature, the spray angle, and the substrate nature have highly significant effects. The materials used as substrate affect splat deformation. A higher thermal conductivity of a substrate results in a lower deformation ratio of splat. The surface tension also affects the resulting shape of the flattened particles. The lower the surface tension, the higher the final diameter. The environment for plasma spraying also affects the adhesion of the coating. The vacuum plasma spraying technique in comparison to the atmospheric plasma spraying technique yields HA coatings characterized by greater adhesion to Ti-6Al-4V.

Microstructure of Coating and Coating Thickness

The internal structure of plasma sprayed coatings is not homogenous and usually consists of grains of different sizes, adhering to each other. The coatings are characterized by a layered structure. Polished cross sections of coatings reveal wavy layers where the outline of each lamella can be distinguished. The macroscopic texture, consisting of a classical lamellar structure and microcracking, is very important in explaining the mechanical behavior of coatings.

For plasma sprayed coatings, to the adhesion substrate decreases with increasing coating thickness. This can be attributed to internal stresses resulting from the difference in the coefficients of thermal expansion between the coating and substrate. Moreover, the coating is being formed of several individual layers deposited in sequence. Internal stresses are also caused by rapid cooling of the coating on the substrate.

Chemical Bonding

Some HA-coated titanium implants were found to have relatively high interfacial shear strength, as measured by using mechanical push-out tests (Thomas, 1987). Chemical bonding may have been generated at the interface, although the majority of the bonding was likely to remain as mechanical interlocking. The HA coating adjacent to the interface was an amorphous phase with a Ca/P molar ratio of about 0.7. In the case of the amorphous phase with its high phosphorous content at the interface, there could be a tendency to form titanium phosphorous compounds as phosphate ions diffuse into titanium, and the amorphous phase could be transformed into the crystalline calcium phosphate with a lower phosphorous content. The reaction products produced at the interface by this interaction could be α-$Ca_3(PO_4)_2$ and $CaTi_2O_5$ (Huaxia, 1992).

1.4.1.6 Bond Strength of Plasma Sprayed Hydroxyapatite Coatings

As for HA coatings, there is a wide variation in the reported bond strength. In the case of air plasma sprayed HA coatings on Ti-6Al-4V, the tensile bond strength was reported to be about 10 – 20 MPa, depending on the coating thickness (Berndt and Gross, 1992). However, bond strength as high as ~60 MPa was also obtained (Wolke et al., 1990). Table 1.12 lists some bond strength values obtained by different researchers.

Table 1.12 Bond strength of plasma sprayed hydroxyapatite coatings

Testing Mode	Substrate	Particle Size of Feedstock /μm	Coating Thickness /μm	Bond Strength /MPa	Source
Tensile	Ti6Al4V	25 - 95	NM **	~10	Cheang and Khor, 1995
Tensile	Ti6Al4V	20 - 45	60 - 80	16.6	Khor et al., 1997
Tensile	Ti6Al4V	75 - 125	< 100	9.4	Khor et al., 1997
Tensile	Ti6Al4V	44 - 125	< 100	52.0 ± 11.7	Lin et al., 1994
Tensile	Ti6Al4V	1 - 45	40	66.8	Oguchi, 1992
Tensile	Ti6Al4V	1 - 45	80	60.7	Oguchi, 1992
Tensile	Ti6Al4V	1 - 45	120	45.3	Oguchi, 1992
Tensile *	Ti6Al4V	~30	140	48.7	Brossa et al., 1993
Tensile	Ti6Al4V	NM **	50	40 - 60	Kay, 1990
Shear *	Ti6Al4V	~30	~88	23.4	Brossa et al., 1993
Shear *	Ti6Al4V	~30	108	20.4	Brossa et al., 1993
Shear	Ti6Al4V	NM **	50	10 - 20	Kay, 1990
Tensile	316 s. s.	25 - 95	NM **	~9	Cheang and Khor, 1995
Tensile	316 s. s.	8 - 40	~50	30	Liu et al., 1994

* Nonstandard testing method; ** Not mentioned

The individual bond type may change, depending on the coating and substrate materials, their properties and the spraying parameters. Because the time of interaction between the ceramic coating and the substrate at high temperature is so short, mechanical anchoring plays an important role in the process of bond formation. The adhesion of ceramic coatings on metal substrates is always poorer than the metallic bond formed while metal is coated on a metal substrate.

Plasma sprayed coatings have been used in many fields. However, the relatively low adhesive strengths lead to limited usage of these coatings. One method to combat poor coating adhesion is to introduce a macrotexture (e. g., steps or grooves) in the substrate to minimize the development of shear stresses along the coating - substrate interface (Thomas, 1987). Another approach is to plasma spray dense HA coatings on a porous substrate (Cook, 1988).

1.4.1.7 Use of Calcium Phosphate Coated Implants

The use of HA coating is mainly on metallic dental endosseous and subperiosteal implants, and to a lesser extent, on metallic orthopedic devices such as total hip and knee prostheses. There are currently very few commercially available HA coated ceramic or polymeric implant devices. On dental and orthopedic implants that have plasma sprayed HA coatings, the coating itself covers only a portion of the device. In a one or two stage root form dental implant, only the portion to be placed within the bone is HA coated, with the exception of some devices in which the coating extends onto the neck, i. e. the transgingival area of the implant. Subperiosteal dental implants are usually entirely coated except for the posts to which the superstructure (e. g., full denture) will be attached. For orthopedic implants, generally only a portion of the contact surface between the implant and bone is HA coated. In some total hip prosthesis designs, only the proximal end of the shaft portion is HA coated, based on the premise that some mobility of the distal end is desirable.

Reports from the early use of HA dental implants by practitioners and the results of some animal studies indicate that plasma sprayed HA coatings cause faster adaptation of bone to an implant device than do uncoated titanium or other implant alloys, although the long-term advantages of the use of these coatings in either dental or orthopedic devices have not been fully established. The rapidly increasing popularity of HA coatings, especially for endosseous dental implants, is perhaps due to the perception that an implant coated with a substance which is similar to bone should result in a more optimum response from either the surrounding bone or soft tissue. In this respect, HA coatings continue to interest both clinicians and researchers as work continues to improve both the design of implants which utilize HA coatings and to optimize such factors as crystallinity and the ion release rate to attain the proper tissue response in each particular orthopedic or dental application (Hench and Wilson, 1993).

1.4.2 Calcium Phosphate Coatings by Ion Beam Sputter Deposition

Due to their excellent mechanical and biological properties, HA coated titanium and its alloys are now used for various applications. Currently, commercially available coated metal implants are manufactured by using the plasma spray technique for depositing HA coatings. However, problems such as relatively low bond strength between the coating and the substrate and non-uniformity across the thickness of the coating are often encountered with plasma sprayed coatings. A variety of other coating techniques, such as electrochemical deposition (Monma, 1993), radio-frequency magnetron sputtering (Wolke et al., 1997), excimer laser deposition (Singh et al., 1994), pulsed laser deposition (Wang et al., 1997), and dipping (Wen et al., 1998), have also been investigated for producing HA coatings on metallic substrates.

In addition to bioactivity, a satisfactory calcium phosphate coating must be dense, hard, adherent and tough for clinical uses. Ion beam sputter deposition was investigated as a method for producing biocompatible ceramic coatings on metallic implants due to its advantages that include the production of thin coatings with high density and superior adhesion (Ong et al., 1992). In this process, ionized argon gas was used to sputter atoms from a ceramic target. The sputtered atoms built up on the metallic substrate that was placed in the path of the sputtered material. Either a argon or nitrogen ion beam could be used for the ion beam bombardment. This technique and its variations are suitable for producing thin bioactive ceramic coatings.

In one series of investigations, the ion beam sputtering/mixing deposition technique was employed to produce thin calcium phosphate (Ca – P) coatings on titanium substrate, and the structure and *in vitro* properties of these coatings were studied (Wang et al., 2000a, 2000b). Commercially available pure titanium was used as the substrate. HA powder was synthesized in-house using the wet method. HA powder disks, which were to be used as the ion beam sputtering target, were uniaxially cold pressed and sintered at 1150℃. The ion beam sputtering/mixing deposition system mainly consisted of one Kaufman ion source and one Freeman ion source, one target holder and one rotatable sample holder in the path of both ion beams. Prior to deposition, etching substrates with argon ions was performed to clean the substrate surface. Monolayer calcium phosphate coatings were sputtering deposited by an Ar^+ beam. Either an Ar^+ beam or a N^+ beam was used to homogenize the coatings. It was found that the Ca/P molar ratio of coatings, as was determined by semiquantitative EDX analyses, varied between 2.0 and 3.0. XRD spectra of Ca – P coatings only exhibited a broad bump and no peaks other than those of titanium substrate were observed, indicating that as-deposited coatings were amorphous. In the Fourier transform infrared spectroscopy (FTIR) spectra obtained, the absorption peaks of PO_4^{3-} were present. No distinct absorption peaks of the hydroxyl group were observed. Using scanning electron microscopy (SEM) and TEM, no structure difference was found in Ca – P coatings produced by the Ar^+ beam and N^+ beam mixing processes. At the microstructure level, the coatings produced were dense, uniform, and adherent. In the dissolution tests of coatings using a physiological saline solution, measurements of ion concentrations in the solution indicated that Ca – P coatings produced by N^+ beam mixing were more soluble than Ca – P coatings produced by Ar^+ beam mixing. Even though no major morphological difference was noted for osteoblast cells on different coatings during *in vitro* tests, the cell growth rate was significantly higher on Ca – P coatings produced by Ar^+ beam mixing than on Ca – P coatings produced by N^+ beam mixing.

1.4.3 Bioactive Functionally Graded Coatings

HA is a brittle and relatively weak bioactive ceramic. Some tough but bioinert ceramics such as TiO_2, Al_2O_3, and ZrO_2 can be used to form bioactive ceramic composites for improved mechanical performance (Salomoni et al., 1994; Champion et al., 1996; Silva and Domingues, 1997). Other calcium phosphates such as TCP can also be considered for toughening HA (Slosarczyk and Bialoskorski, 1998). Additionally, due to its resorbability, TCP in the HA-based composites can gradually dissolve after implantation, so that enhanced osseointegration is promoted.

One method to improve the properties of coatings and the bonding between the coating and the substrate is to employ a coating with a novel structure. A functionally graded material, which has a compositional gradient from the surface to the interior of the material, can be used for such purposes. To strike a balance between mechanical properties and biocompatibility, a three-layer functionally graded coating (FGC) structure may be used. The sublayer which bonds with the substrate should be a mechanical support for the whole coating system. In the meantime, the sublayer on the surface of the FGC should be bioactive (or bioresorbable) in order to obtain good biocompatibility with osseous tissues. The sublayer that is located between the aforementioned layers may have a compromise mechanical/biological performance.

Several bioactive FGCs were made using the plasma spray technique according to the adhesive strength and bioactivity (or bioresorbability) of various components in the coatings (Khor et al., 1997). The structure and properties of these FGCs were subsequently studied (Wang et al., 1999). The flame spheroidized HA (SHA) used in the investigation was produced in-house. Commercially available α-TCP and TiO_2 were mixed with SHA according to predetermined compositions of sublayers in FGCs. Two types of FGCs were obtained by using (a) mixtures of SHA and spherical α-TCP powders and (b) mixtures of SHA and TiO_2 powders. The sublayer adjacent to the Ti-6Al-4V substrate was either SHA or TiO_2. FGCs were produced by plasma spraying the ceramic mixtures layer by layer under constant spraying conditions. The thickness was around 130 μm for each sublayer in FGCs. The as-sprayed coatings were dense and in the sublayers the microstructure was composed of randomly stacked lamellae. XRD patterns of coatings showed that, apart from HA, the coatings also contained some other phases such as CaO and TTCP. The appearance of these new phases was due to the decomposition of HA caused by the high temperature within the plasma flame. Peak heights of the HA phase decreased and its width became broad due to the formation of the amorphous phase during the plasma spray process. Using microindentation tests on the cross section of FGCs, variations in the mechanical properties in the FGCs were clearly shown. In the SHA – TiO_2 FGC, the first sublayer which was composed of TiO_2 exhibited a high microhardness value (around 940 VHN). The microhardness decreased dramatically to 250 VHN in the second layer which had 50 wt% of SHA. The microhardness value of the third layer (i. e. the SHA layer) was comparable to the value obtained for the monolayer SHA coating. It was found that the microhardness in the transitional area between adjacent layers was relatively low. These low values were attributed to the relatively high residential stress at these locations. The indentation fracture toughness was also calculated for each sublayer of the FGCs, and it was shown that the addition of tougher ceramics (either α-TCP or TiO_2) could improve the fracture toughness of pure HA coatings significantly, with TiO_2 offering greater toughening efficiency.

To overcome the problem of HA decomposition due to high temperatures in the plasma spraying process, an FGC consisting of HA and TCP was successfully produced using the spray-and-sinter technique (Kumar et al., 2000c). Commercially available spherical HA (SHA) and α-TCP were used to make HA/TCP composite powders. The composite powders were produced by mixing appropriate amounts of HA (e. g. 50 wt%) and TCP (e. g. 50 wt%) in a laboratory planetary mill. The FGC was subsequently manufactured by spraying composite powders of different compositions layer by layer onto Ti-6Al-4V plates at room temperature using an airbrush under controlled pressure. The FGC on Ti-6Al-4V was sintered at 900℃. Individual sublayers of the five-layered FGC contained 100% SHA, 75% SHA + 25% TCP, 50% SHA + 50% TCP, 25% SHA + 75% TCP, and 100% TCP, respectively. XRD analysis revealed that no HA decomposition was encountered with this coating technique and the FGC retained functional groups for its bioactivity, as shown in FTIR spectra. Using the nanoindentation technique, Young's modulus and the hardness of the 50% SHA + 50% TCP layer were determined to be 15.4 GPa and 0.68 GPa, respectively.

1.5 Bioactive Composites

1.5.1 Bioactive Composites for Tissue Replacement and Regeneration

Various engineering materials, including stainless steel, cobalt – chromium alloys, titanium alloys, alumina, and zirconia, have been used in orthopedic surgery as replacements for bone. However, these materials possess much higher moduli than that of cortical bone (Table 1.13). According to "Wolff's Law", bone will remodel in a new mechanical environment so that the stress or strain is retained within specific levels (Wolff, 1892). If an implant which is stiffer than bone is placed in bone, bone will be subjected to a reduced mechanical load and will hence resorb (Huiskes et al., 1992).

The development of analogue biomaterials is based on the concept that the best material for replacing a body tissue is one that is similar, if not identical, to that tissue. If similar biological and mechanical properties can be achieved through a new implant material, then the stress concentrations at the interface between the implant and the tissue will be minimized, as will be the

Table 1.13 Mechanical properties of current implant materials and skeletal tissues

Material	E/GPa	σ/MPa	ε/%	K_{IC}/MN · $m^{-3/2}$
Alumina	400	450	~0.5	~3
Co – Cr alloys	230	900 – 1540	10 – 30	~100
Austenitic stainless steel	200	540 – 1000	6 – 70	~100
Ti-6Al-4V alloy	106	900	12.5	~80
HDPE	1	30	>300	—
Cortical bone	7 – 30	50 – 150	1 – 3	2 – 12
Cancellous bone	0.05 – 0.5	10 – 20	5 – 7	—
Articular cartilage	0.001 – 0.01	10 – 40	15 – 50	—
Tendon	1	80 – 120	10	—

E: Young's modulus; ε: elongation at fracture; σ: tensile strength (in the case of alumina: flexural strength); K_{IC}: fracture toughness

disruption of the associated cellular activity of the body. Bone itself is a nanoscale composite consisting of bone mineral particles approximately 5 nm × 5 nm × 25 nm in size, which are calcium-deficient, substituted hydroxyapatite in collagen, a natural polymer. Therefore, a polymeric matrix composite containing a bioactive phase appears a natural choice to substitute for bone. In the early 1980s, Bonfield and his co-workers pioneered research into bone analogue materials and produced a bioactive composite with hydroxyapatite (HA) particles dispersed in high density polyethylene (HDPE) (Bonfield et al., 1984). Attempts were made later to use collagen itself as a matrix for bioactive composites (Clarke et al., 1993). But the modulus of demineralized collagen was low (0.34 GPa), and only a small amount of HA could be incorporated, giving an increase in Young's modulus to 0.76 GPa which is too low for bone substitution. Since its invention, the hydroxyapatite reinforced polyethylene composite (now designated as HAPEX™) has been extensively characterized in terms of its mechanical properties and its *in vitro* and *in vivo* biological response. Based on the same technology, new bioactive composites have also been produced and assessed. The design of an appropriate combination of mechanical and biological properties for bone replacement materials has proved attractive to clinicians. The superior fracture toughness and trimmability of polymeric matrix composites compared to existing monolithic bioactive ceramics allow the implants to be shaped by surgeons during operations. These attributes can lead to wide clinical use of bioactive composites.

1.5.1.1 HAPEX™

Polyethylene is a proven biocompatible material which is widely used in orthopedics while hydroxyapatite closely resembles bone mineral. It was therefore reasonable to combine the two materials to produce a composite which would match cortical bone, both mechanically and biologically. Ductile polyethylene allows the incorporation of a high percentage of bioactive particles in the polymer which is essential for achieving bioactivity in the composite. Furthermore, with advanced processing technology, the mechanical properties of the composite can be significantly enhanced, enabling surgeons to use the material in different areas for low-load and major load-bearing applications.

Initially, calcined bone ash (CBA) was used as the bioactive phase (Bonfield et al., 1981). Hence, the processing route and technological parameters were optimized using the CBA - HDPE system (Abram et al., 1984). Later, synthetic HA particles were used and this synthetic HA-HDPE composite has become known as HAPEX™ (Wang et al., 1994). In addition to tensile mechanical properties, dynamic (Wang et al., 1998a), creep (Suwanprateeb et al., 1995), and fatigue (Tanner et al., 2000) properties have also been investigated. It was found that the mechanical properties of HAPEX™ could be altered by varying the mean particle size and particle size distribution of HA and the polyethylene matrix (Wang et al., 1998b). In a rheological investigation, HAPEX™ exhibited discontinuity with a varying shear rate at processing temperatures (Tang et al., 2000). The discontinuity was attributed to the building up and sudden yielding of elastic axial forces under sufficient stress. As the HA content increased, the shear rates at which discontinuity occurred decreased. A detailed analysis of the rheological behavior of HAPEX™ has helped to select optimal processing conditions.

The manufacture of HAPEX™ consists of blending, compounding, centrifugal milling to produce a composite powder, and compression or injection molding to produce bulk material or devices. Composites with up to 45 vol% (i.e. 73 wt%) of HA can be routinely made with HA particles evenly distributed in the polyethylene. Uniform distribution of the bioactive filler in a composite is essential for the mechanical performance of the materials. By varying the amount of hydroxyapatite in HAPEX™, a range of mechanical properties of the material (Table 1.14) and biological response to the material can be obtained. An increase in the HA volume percentage increases Young's modulus of HAPEX™ but decreases the strain to fracture. A ductile - brittle transition exists at about 40 vol% of HA. HAPEX™ with 45 vol% of HA has a Young's modulus of 5.54 GPa, which approaches the lower bound for cortical bone. Even with such a high percentage of the brittle phase, the composite can still be readily shaped with a scalpel in the operating theater, enabling the implant to be made an exact fit for the patient, thus being "surgeon friendly" (Tanner et al., 1994). It was found, through *in vivo* studies, that a minimum amount of HA (approximately 20 vol%) is required in the composite for bone apposition and bone bonding to occur in a mechanically lightly loaded implant (Bonfield et al., 1986).

The processing route adopted results in only a mechanical interlock between

Table 1.14 Mechanical properties of HAPEX™, Bioglass®/HDPE and A-W glass-ceramic/HDPE composites

Volume of Reinforcement/%	Young's Modulus/GPa			Tensile Strength/MPa			Elongation at Fracture/%		
	(1)	(2)	(3)	(1)	(2)	(3)	(1)	(2)	(3)
0	0.65	0.65	0.65	17.89	17.89	17.89	>360	>360	>360
10	0.98	1.05	0.96	17.30	14.34	17.32	>200	105	>180
20	1.60	1.21	1.34	17.77	12.69	16.67	34	64	130
30	2.73	—	1.83	19.55	—	14.68	6.40	—	28.70
40	4.29	2.54	2.84	20.67	10.15	14.87	2.60	8.50	5.30

(1) HAPEX™; (2) Bioglass®/HDPE; (3) A-W glass-ceramic/HDPE

HA and HDPE in HAPEX™, without any chemical bonding. The mechanical coupling has resulted from shrinkage of the HDPE matrix around individual HA particles during cooling from processing temperatures. Examination of fracture surfaces of HAPEX™ after tensile and creep tests has shown that HA particles debonded from the matrix polymer during testing, forming dimples in the composite. This failure mechanism indicates that a relatively weak bond exists at the reinforcement-matrix interface (Friedrich and Karsch, 1981). Such an interface may allow water uptake and hence can lead to reduced mechanical properties when the material is subjected to an aqueous environment (Suwanprateeb et al., 1997).

Predictive modeling of composite materials allows an understanding of their mechanical behavior and enables the effects of the properties of constituent materials to be explored. Establishing an effective model for the finite element (FE) analysis of particulate filled polymers, including HAPEX™, can greatly assist in developing new generations of bioactive composites. The FE model itself had been successfully used for particulate filled epoxy resin (Guild and Young, 1989a, 1989b) and was applied to the hydroxyapatite – polyethylene system to predict the moduli of HAPEX™ (Guild and Bonfield, 1993). Using simple assumptions of linear elasticity and perfect bonding between the reinforcement and the matrix, the variation in Young's modulus of HAPEX™ with HA volume fraction could be predicted. Both the experimental and predicted values of the modulus increased with increasing HA volume fraction. There was good agreement, particularly after the application of the statistical model, between experimental and predicted values except at high HA volume fractions (Guild and Bonfield, 1993). In addition to Young's modulus, bulk and shear moduli of HAPEX™ can also be predicted using this model, which assists in assessing the suitability of HAPEX™ for certain applications.

1.5.1.2 Chemically Treated HAPEX™

The polymer structure and molecular mass, the filler type and volume fraction, and the interfacial state between the filler and the polymer affect the properties of particulate filled polymers. The interface can be modified by filler surface treatment and/or modification of the matrix polymer. Various research groups introduced an interfacial phase between the reinforcement and the matrix for polymer composites (Xavier et al., 1990; Maiti and Sharma, 1992; Fu and Wang, 1992). In general, an interfacial phase promotes adhesion between the reinforcement and the matrix and hence improves the mechanical properties of composites.

Polyethylene is a nonpolar, hydrophobic polymer, and consequently in HAPEX™, only mechanical interlock exists between the HA particles and the HDPE matrix. In order to achieve good affinity between these two phases, it is important for the polymer to bear some polarity. On the other hand, silanation has proved to be an effective method for treating HA particles for reinforcing a bone cement (Khorasani et al., 1992). Therefore, both silanation of HA and acrylic acid grafting of HDPE were investigated as methods of improving reinforcement – matrix coupling for HAPEX™ (Wang et al., 1996a). The silane coupling agent was 3-trimethoxysilylpropylmethacrylate, and analytical grade acrylic acid was used for grafting. Infrared spectra obtained from chemically coupled HAPEX™ exhibited spectral peaks arising from the acrylic acid and the silane coupling agent. EDX analysis of fracture surfaces obtained at liquid nitrogen temperature indicated a silicon-containing interphase between HA and polyethylene in chemically coupled HAPEX™. By acrylic acid grafting and silanation of HA, a synergistic effect was obtained as a result of improved adhesion and interaction between the filler and the matrix. Therefore, up to a 12% increase in strength was achieved for chemically coupled HAPEX™ through both silanation and acrylic acid grafting. The fracture mechanism for chemically coupled HAPEX™ was not fundamentally different from that for the original HAPEX™. However, with the introduction of silanated HA and acrylic acid grafted polyethylene, a stronger bond (both chemical and mechanical) formed between the filler and the matrix, which delayed the initial debonding and cavitation process, and thus effectively increased the ductility of the composite.

1.5.1.3 Hydrostatically Extruded HAPEX™

For major load-bearing applications, it is necessary to increase substantially the stiffness and strength of HAPEX™. Chemical coupling of HA to polyethylene has resulted in improved mechanical behavior but only limited increases in mechanical properties. More significant levels of improvement, however, are required. Molecular orientation in a polymer leads to a significant enhancement in the stiffness and strength in the orientation direction (Frank, 1970). These effects are particularly evident with polyethylene due to its simple molecular structure (Capaccio and Ward, 1974; Gibson et al., 1974). Among several well-established technologies for inducing molecular orientation in polymeric materials, hydrostatic extrusion has been used to increase substantially the tensile strength and modulus of polyethylene (Gibson et al., 1974), It is therefore natural to investigate the possibility of hydrostatically extruding HAPEX™ (Wang et al., 1997).

It appeared that hydrostatic extrusion did not alter the distribution of HA in HAPEX™ which was homogeneous after compounding. A solid-state deformation process, such as hydrostatic extrusion, does not introduce any interfacial reaction between HA and polyethylene during manufacture. Therefore, as in the original HAPEX™, there was only mechanical interlock at the interface in hydrostatically extruded HAPEX™. However, such an interlock may become stronger after hydrostatic extrusion, because the high shear stresses employed during the extrusion process could enhance the penetration of polymer into particles, thus improving the mechanical bond. It was found that higher extrusion ratios led to higher a Young's modulus and tensile strength of HAPEX™ (Table 1.15). The fracture strain of HAPEX™ was also substantially increased by hydrostatic extrusion. At an extrusion ratio of 8:1, Young's modulus and the tensile strength of HAPEX™ with 40 vol% of HA were 2.6 and 4.4 times higher than those of compression molded material. This material also possessed a strain to fracture which is far greater than that of human cortical bone (9.4% vs. 0.5% – 3.0%). Therefore, HAPEX™ further processed via hydrostatic extrusion shows great potential for major load bearing applications.

Table 1.15 Mechanical properties of hydrostatically extruded HAPEX™ (V_{HA} = 40 vol%)

Extrusion Ratio	Young's Modulus /GPa	Tensile Strength /MPa
1:1	4.29	20.67
5:1	5.89	64.78
8:1	9.91	91.23

Chopped polyethylene fibers have also been used as a matrix in the hydrostatic extrusion of HA reinforced polyethylene (Ladizesky et al., 1996). The polyethylene fiber possesses much higher strength and modulus values than conventionally processed polyethylene since the polymer chains are already aligned along the fiber axis. Using these fibers as the matrix should give much higher mechanical properties in composites. The HA particles and chopped polyethylene fibers were first dry-mixed and then compression molded into a large diameter composite rod. The molding temperature was chosen so as to melt only the sheaths of the polyethylene fibers. Fibers were nearly aligned in the extrusion direction after hydrostatic extrusion, although their length seemed to be significantly shortened. In such composites, HA particles appeared as colonies surrounded by the fibers. Hydrostatically extruded fiber matrix composites possessed the highest strength and moduli in the hydroxyapatite – polyethylene system.

1.5.1.4 Bioglass® Reinforced Polyethylene

The utilization of Bioglass® as a biomaterial in its own right, a coating material, a matrix, or as a toughening, secondary phase in a ceramic has created a range of new bioactive materials. A particular advantage of Bioglass® is the ability to bond to both hard and soft tissues. However, as for all monolithic bioactive glasses and ceramics, the disadvantage of Bioglass® is that it is a brittle solid and, as a consequence, its handling and mechanical properties are not adequate for load-bearing applications. Experience with HAPEX™ has demonstrated that appropriate mechanical properties, including ductility, can be developed for a composite, while retaining a bioactive response *in vivo*. Therefore, the manufacturing technology developed for HAPEX™ was used to produce Bioglass® reinforced polyethylene composites, which are expected to offer a distinctive combination of mechanical/biological performance and hence some potential clinical applications (Wang et al., 1995).

Bioglass® particles were found to be well dispersed, and a reasonably homogeneous distribution of particles in the polymer matrix was achieved after the compounding process (Fig. 1.14). Such a distribution of Bioglass® particles is important for both the mechanical behavior of the composite and the biological response to the composite. There was no chemical bond between the Bioglass® particles and the polymer matrix, which accounted for the decrease in composite strength with increasing Bioglass® content (Table 1.14). Composites with up to 30 vol% of Bioglass® exhibited elastic compliance, tensile strength, and fracture strain comparable to those of soft connective tissues. Composites with Bioglass® volumes in excess of 30% displayed mechanical properties comparable to cancellous bone.

Figure 1.14 A homogeneous distribution of Bioglass® particles in a Bioglass®/HDPE composite containing 40 vol% of Bioglass®

The incorporation of particulate Bioglass® into polyethylene has resulted in a composite with a higher modulus than the unfilled polyethylene. However, these modulus values were consistently lower than those of HAPEX™ with corresponding HA volumes, although a similar ascending trend was observed with an increasing amount of Bioglass® in the composite. Such a difference in modulus values can be attributed to the differences in mean particle size, particle size distribution, and mechanical properties of HA and Bioglass®. One of the reasons for the lower strength values obtained from the Bioglass®/HDPE composite is that due to the smoother Bioglass® surface, a weaker mechanical coupling exists between Bioglass® and polyethylene than between HA and polyethylene.

1.5.1.5 A-W Glass-Ceramic Reinforced Polyethylene

A-W glass-ceramic exhibits high bioactivity and has already been used for vertebral prostheses in patients. However, like other bioactive glasses or ceramics, it is brittle and thus has limited applications. Thus, A-W glass-ceramic reinforced polyethylene composite was developed (Wang et al., 1996b), extending the spectrum of available bioactive composites to parallel the major monolithic ceramics (hydroxyapatite), glasses (Bioglass®), and glass-ceramics (A-W glass-ceramic) in clinical use.

As with HAPEX™ and Bioglass®/HDPE composites, a homogeneous distribution of A-W glass-ceramic particles in the polyethylene matrix was achieved after compounding and compression molding. Young's modulus of the composite also increased with an increase in the A-W glass-ceramic volume, while the tensile strength and fracture strain decreased. These observations are consistent with the results obtained from other composites. However, even with 40 vol% of the glass-ceramic, the composite still exhibited considerable ductility. There was no chemical bonding either between the A-W glass-ceramic particles and the polymer in the A-W glass-ceramic/HDPE composite.

1.5.1.6 Hydroxyapatite Reinforced Polysulfone

Instead of using physical means such as hydrostatic extrusion for composites having semicrystalline polymer matrices, another way of increasing the stiffness and strength of bioactive bone analogue materials is through the use of polymers that have relatively high modulus and strength values. Polysulfone (PSU) has properties matching those of light metals. Its favorable properties include high strength and modulus, low creep rate, excellent resistance to hydrolysis or reduction of molecular weight, stability in aqueous acids, alkalis, and salt solutions, and bioinertness. It also has high resistance to β-, γ- and X-ray radiation and can be steam-sterilized. Therefore, the HA/PSU composite has been developed as a new tissue replacement material (Wang et al., 1999). A new manufacturing process was established for HA/PSU, and its processing conditions were investigated. It was also found that the stiffness of the composite increased with an increase in HA content. In an immersion experiment, water uptake reached an equilibrium after 7 days' immersion in distilled water for the HA/PSU composite and unfilled PSU. After 7 days' immersion in distilled water, the storage modulus of the HA/PSU composite decreased less than that of unfilled PSU.

1.5.1.7 Calcium Phosphate Reinforced Polyhydroxybutyrate and its Copolymer

The potential applications of polyhydroxybutyrate (PHB) come from the exploitation of its ability to degrade and resorb. Young's modulus of PHB is relatively high (4 GPa) as compared with HDPE (0.7 GPa). Being a thermoplastic, PHB can be processed by conventional techniques such as extrusion, injection and compression molding, although care must be taken in controlling the processing conditions due to the heat sensitivity of the polymer. HA reinforced PHB was produced using the technology established for HAPEX™ (Doyle et al., 1991). As expected, there was a correlation between the HA content and composite modulus, with the modulus rising from 4 GPa for the molded polymer to 11 GPa for the composite containing 40% HA, while the strength and strain to fracture decreased correspondingly. It was found that PHB was susceptible to hydrolysis when immersed in buffered saline at 37℃. The modulus of a HA/PHB composite containing 20% HA decreased from over 9 GPa to 5 GPa after four months' exposure to saline at 37℃. Bending strengths of unfilled PHB and HA/PHB composite followed the same trend. From quantitative histology, it was noted that statistically significant differences existed between the amount of bone formation induced on the surface of HAPEX™ and that of HA-filled or unfilled PHB. Analysis showed that PHB-based materials produced superior bone healing with greater than double the bone-conducting ability as compared with HA reinforced polyethylene.

Recently, work started on developing tricalcium phosphate (TCP) reinforced PHB and its copolymer (Wang et al., 2000). The processing technology was similar to that used for HA/PSU, and a TCP/PHB composite containing up to 30 vol% of TCP was produced. TCP ceramic particles were well distributed in the polymer. DSC analysis showed that an increase in the TCP content resulted in decreases in both the melting temperature and the crystallinity of PHB. The stiffness of the composite increased with an increase in the bioceramic content. Due to the biodegradability of its constituents, the TCP/PHB composite has a potential for various clinical applications.

1.5.1.8 Calcium Phosphate Reinforced Chitin

Synthetic, poorly crystallized HA is similar to bone apatite, and it has been shown that it has greater ability than fully crystallized HA to induce the formation of a bone-like apatite *in vitro* and is hence more bioactive (Weng et al., 1997). Chitin acting as a structural and defensive material, occurs in crustacea, mollusks, and insects and is an important constituent of the exoskeleton. It is one of the most abundant organic materials, second only to cellulose in the amount produced annually by biosynthesis (Tracey, 1957). Chitin as a natural polymer is biodegradable due to its β-1,4 glycosidic linkages being susceptible to the lysozyme present in the human body (Roberts, 1992). Chitin is therefore a potential biodegradable polymer that can be used for tissue replacement and regeneration. It was expected that the incorporation of poorly crystallized apatite into chitin would produce a composite as an advanced bioactive and biodegradable material. Therefore, using the solution casting technique, a poorly crystallized HA/chitin composite was made as a new material for bone regeneration (Weng and Wang, 2000). After optimizing the processing conditions, poorly crystallized HA particles were found to be evenly distributed in the composite. *In vitro* mineralization experiments showed that these particles rendered the composite bioactive and improved its ability to induce the formation of bone-like apatite on composite surfaces. An incubation temperature higher than 36.5℃ accelerated the nucleation of bone-like apatite on the composite surface because of the dissolution of D-glucopyranose residues available in chitin which led to an increase in hydroxyl concentration in the vicinity of the surface. The alkaline treatment of chitin also introduced more D-glucopyranose in chitin and hence significantly improved the *in vitro* nucleation of bone-like apatite on surfaces of chitin and its composite when they were immersed in simulated body fluid.

1.5.1.9 Biological Response to Bioactive Composites

When a bone analogue material is implanted into the body, the response to the material will depend on a combination of the responses to the constituent phases of the material. The biological performance can be evaluated by *in vitro* tests, using simulated body fluid or human osteoblast cell cultures, or *in vivo*. A characteristic *in vitro* response of HAPEX™ is shown in Fig. 1.15 (Huang et al., 1997), which demonstrates the osteoblast cellular attachment to "islands" of HA and the subsequent cell proliferation. These observations clearly show the bioactivity that can be created on a composite implant surface.

For *in vivo* tests, HAPEX™ was implanted into the lateral femoral condyle as 2.4 mm diameter, 5 mm long pins inserted as a press-fit (Bonfield et al., 1986). Undecalcified histology was performed after 1 month, 3 months and 6 months postimplantation. The implant was initially surrounded by a fibrous layer, which was again surrounded by a layer of new bone. With time, the fibrous layer was replaced by bone contact in some areas. At 6 months, the areas of direct bone apposition, measured from histological sections, were 40% of the implant surface. These histological blocks were then used to prepare ultramicrotomed TEM sections of either fibrous capsule regions or bone – implant contact regions. The sections were stained with uranyl acetate and osmium oxide

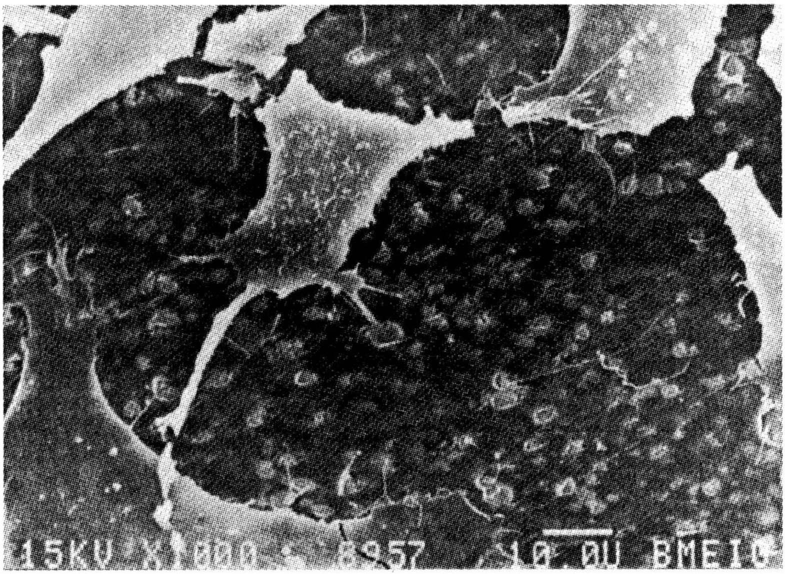

Figure 1. 15 Human osteoblast cells attaching to hydroxyapatite particles in HAPEX™ (cell culture at 24 h)

vapor or left unstained for high resolution TEM (HRTEM). At 1 month, TEM examination showed the interface to be approximately 6 μm thick, but by 6 months, in the regions where it remained, it had reduced to 2 μm thickness. HRTEM showed the atomic lattice of apatite crystals both within the bone and in the composite. In some regions of direct bone – implant contact, the atomic lattices were seen to be continuous across the interface, indicating epitaxial growth.

To further investigate tissue response to HAPEX™, cylindrical implants with a slot cut into the side were implanted into the lateral femoral condyle of rabbits (Revell and Tanner, 1998). They were compared with identical titanium implants with or without an HA coating plasma sprayed only into the slot. Bone growth into the slot was similar in both the HA coated titanium and HAPEX™ slots and was significantly greater than that into the uncoated titanium implants.

1.5.1.10 Clinical Applications

Subperisteal orbital floor implants made from HAPEX™ have been used for patients. When HAPEX™ was put into patients, the implants were inserted below the periosteum, the fibrous layer covering bone, and were merely left in position (Downes et al., 1991; Tanner et al., 1994). The orbital floor has a ridge at the front which helps to prevent movement of the implant. But no sutures or wires were used to stabilize the implant position. Computer tomography was unable to distinguish an interface between the implant and the bone after 6 months implantation. Clinical examination found that the implants felt stable. In one patient where subsequent surgery was needed for unrelated reasons, the implant was found to be totally stable.

HA implants have been used increasingly over the past 10 years to reconstruct the ossicular chain. Because of technical difficulty encountered in the precise shaping of dense HA, many implants are now constructed in a two-piece design featuring an HA head and a trimmable shaft. Recently, HAPEX™ has been used as a middle ear implant material, and various designs of ossicular replacement prosthesis using HAPEX™ have appeared (Swain et al., 1999). In HAPEX™, the presence of HA (40 vol%) provides bioactivity to the material as its means of achieving enduring implant stability. HAPEX™ also possesses adequate mechanical and physical properties and can be "custom" trimmed intraoperatively. These prominent features have resulted in the refinement of middle ear implants and should lead to improved designs in the future. Preliminary clinical results were satisfactory, and there was no case of extrusion of HAPEX™ implants. Total HAPEX™ implants can be produced by injection molding, which will result in significant reductions in the manufacturing costs of these implants.

1.5.2 Bioactive Bone Cement

In joint replacement surgery, there are two main methods of fixation of the prosthesis: cemented or uncemented. The most common method chosen to secure prostheses, particularly in patients over 60 years of age, is the use of polymethylmethacrylate (PMMA) based bone cement. Preformed PMMA beads are mixed with methylmethacrylate (MMA) monomer. The polymerization process occurs as a result of the reaction between the benzoyl peroxide in the polymer powder and N,N-dimethyl-*p*-toluidine in the monomer. In practice, the cement is mixed in surgery until it becomes a doughy consistency when it can be inserted into the bone cavity, and the completion of the polymerization takes place *in vivo*. Although joint replacements are generally very successful, failures of prostheses can occur. These failures are mainly due to aseptic loosening which is often caused by the failure of the cement mantle (Gruen et al., 1979). A good interface between the cement and bone is very important. In reality, due to the high exotherm of the polymerization reaction and the toxicity of the MMA, bone necrosis can occur. Consequently, there is fibrous encapsulation around the implant which allows micromotion to occur, causing pain to the patient and a space for wear particles to accumulate (Anthony et al., 1990). A solution to overcome this problem at the interface is to use a bioactive bone cement. It has been suggested that three types of bioactive bone cements can be used (Oonishi, 1991):

(1) An all-bioactive bone cement where the whole material is bioactive, for example, calcium phosphate cements.

(2) A surface-bioactive cement where bioactive filler particles are added to a nonbioactive matrix, for example, HA added to PMMA cement.

(3) An interface-bioactive bone cement where a bioactive material is placed between the bone and nonbioactive cement, for example, a layer of HA granules between bone and cement.

Extensive studies have been conducted on the second type of bioactive bone cement.

Currently, all commercial bone cements are based upon PMMA and/or MMA copolymers. A variety of filler particles have been used for increased bioactivity and for improving the relatively poor mechanical characteristics of bone cements. These bioactive fillers include hydroxyapatite and bone particles. Growth hormone can also be added.

Even though its mechanical properties are not ideal, in terms of bioactivity, HA is a good choice as a reinforcing particle in a bioactive bone cement. The addition of up to 15 wt% of HA to a PMMA cement was investigated, and mechanical and setting characteristics were examined (Olmi et al., 1983; Castaldini and Cavallini, 1986). The highest Young's modulus (2.5 - 3.5 GPa) and flexural strength (58 - 69 MPa) were obtained by adding 3 wt% of HA. Bending creep tests showed an improvement in creep resistance with 5 wt% of HA. Furthermore, the maximum temperature during polymerization was reduced by approximately 20% with 5 wt% of HA which was believed to be due to the HA filler acting as an internal heat sink in the polymeric matrix. The influence of the addition of HA powder with different average particle sizes was assessed via both static and dynamic mechanical tests (Sogal and Hulbert, 1992; Dove and Hulbert, 1995). Static tests revealed that up to 10 wt% of HA particles with an average particle size of 96 μm could be added without any large decreases in tensile strength. The inclusion of 20 wt% of HA produced a decrease in strength with all sizes of HA particles. Fatigue results showed that adding 10 wt% of HA particles of different average sizes had no significant effect or actually increased the fatigue resistance of the cement composite. The inclusion of 2 wt% - 8 wt% HA fibers into a bone cement was found to increase the elastic modulus, compressive strength, and fracture toughness, whereas creep the resistance and maximum curing temperature of the cement composite decreased (Murakami et al., 1992). Unfortunately, there was a limited amount of HA fiber which could be added to PMMA. Even though it is possible to include certain amounts of HA either as a powder or fiber into PMMA bone cement, the amount which can be included while maintaining mechanical strength or handling properties of composite cement may be so small that the increase in bioactivity is not likely to be very large.

Bone particles are another source of bioactive filler material. The mechanical properties of a low-viscosity cement mixed with particles of bovine tibial cancellous bone were assessed (Williams and Johnson, 1989). Results showed that, with up to 25 vol% of bone particle, Young's modulus increased from 2.4 GPa to 4.4 GPa. Both the mechanical and biological characteristics of an inorganic bone particle impregnated PMMA cement were also studied with promising results (Park et al., 1986; Henrich et al., 1993). The incorporation of up to 30 wt% of bone particles was shown to improve the fatigue resistance of the reinforced cement with no decrease in tensile strength. An *in vivo* weight bearing investigation in dogs showed that after 1 month of implantation, active bone growth into the cement had occurred and this continued for up to 5 months. By this time, the majority of the surface bone particles had been replaced by new bone. Push-out tests supported this observation, revealing the shear strength of the bone particle impregnated cement-bone interface to be 3.6-fold that of the control cement – bone interface. These studies show clearly the potential of a bone cement with increased bioactivity.

A different approach to increasing the bioactivity of PMMA bone cement is the use of human growth hormone (Downes et al., 1990; Goodwin et al., 1997). *In vivo* tests were conducted using a New Zealand white rabbit model. Results from quantitative histomorphometric analysis revealed that 1 month after implantation, there was a greater percentage of osteoid and mineralized bone at the surface of the hormone loaded cement than at the surface of the plain cement. Four months after implantation, there was much less difference between the two cements. This was reflected by drug release tests, where the major fraction of hormone was released within 24 hours from the start of testing. There can be some early benefit to the addition of hormone to bone cement. To produce long-term effects, slower drug release is desirable.

Alternatives to PMMA cement as a polymeric matrix have also been investigated in order to overcome the brittleness problem of PMMA. Bioactive cements using bisphenol-*a*-glycidyl methacrylate (BIS-GMA) or polyethylmethacrylate/*n*-butylmethacrylate (PEMA/*n*BMA) as the matrices were developed (Kawanabe et al., 1993; Behiri et al., 1991).

1.5.3 Bioactive Ceramic Matrix Composites

1.5.3.1 Glass Toughened Hydroxyapatite

Hydroxyapatite in solid form has been successfully used as a space filler and for bone grafts (Gatti et al., 1990; Inoue et al., 1991, 1992; Shinjo et al., 1991). However, it is limited in its use because its mechanical properties are not adequate for load-bearing applications such as spinal fusion. Various techniques have been investigated to improve the mechanical properties of HA. An obvious approach is to toughen HA with glass. Glass can exist as a liquid phase at the grain boundary during sintering and toughen the ceramic after sintering. This method of liquid-state sintering has been used successfully with engineering ceramics such as silicon carbide and silicon nitride (Riley, 1983).

A glass reinforced HA was developed with significantly enhanced mechanical properties compared to unreinforced HA (Santos et al., 1992). Increases in both the flexural strength and the fracture toughness were between threefold and fourfold. However, the improvements in mechanical properties could be at the cost of the stability of the phases present due to the inherent instability of the HA and the role of glass to usually promote HA transformation to TCP at sintering temperatures.

The effect of individual oxides in the reinforcing glass in promoting phase changes and hence affecting mechanical properties was investigated in detail (Knowles and Bonfield, 1993). Two glass systems were used: (a) the $CaO - P_2O_5$ system, due to its close relationship to the chemical composition of HA; and (b) the $Na_2O - P_2O_5$ system, as the presence of soda in a glass when combined with HA is known to promote premature phase changes. Glasses containing various amounts of the network modifying oxide (soda and lime) were used, and the overall glass content in the glass - HA composite was varied. With the lime glass at 2.5 wt% addition, a similar pattern of phase changes was seen irrespective of glass composition. As the sintering temperature increased from 1200℃ to 1350℃, an increase in the β-TCP content was detected using XRD analysis. A concurrent β-TCP to α-TCP transformation also occurred. Although the flexural strength was significantly increased, no significant trends for the effect of glass composition on flexural strength could be seen. The results for soda glass were different from those for lime glass. The soda produced considerable phase changes within the HA. Analysis of the phases present when a 2.5 wt% glass addition was used, irrespective of composition, showed that it gave rise to β-TCP at temperatures as low as 1200℃. With sintering temperatures above 1250℃, α-TCP also appeared. As the soda content in the glass increased, the inversion of β-TCP to α-TCP with increasing sintering temperature became reduced. Therefore, it seemed that the soda was stabilizing the β-TCP phase. The effect of the stabilization of β-TCP at high temperatures, which was seen with a 2.5 wt% glass addition, became even more apparent at 5 wt% addition. The mechanical properties clearly reflected the phase changes, with a rapid drop in flexural strength for all glass types above 1250℃. The stabilization of β-TCP, i.e. hindrance of the formation of α-TCP, was reflected in the flexural strength. The lower the percentage of α-TCP formed, the better the mechanical properties.

It appears that to produce a reinforced HA, the use of a $CaO - P_2O_5$ type glass is of significant benefit to both the stabilization of the HA and the production of high flexural strength. Furthermore, the amount of glass added must be of the order of 2.5 wt% to minimize the reactive liquid phase present at the grain boundaries, which should still be present in sufficient quantities to sinter the powder.

1.5.3.2 Metal Toughened Hydroxyapatite

A bioactive glass-ceramic incorporating ductile metal particles was produced and assessed (Taylor, 1994; Claxton et al., 1996). The matrix of the composite was Apoceram which is a glass-ceramic containing apatite and wollastonite. Small amounts of sodium silicate and up to 20 vol% of particulate titanium were added to make the composite. A small addition of sodium silicate allowed the composite to be fabricated at a lower temperature than for Apoceram. It was found that the flexural strength of the composite was significantly improved as compared with the parent Apoceram, even though the fracture toughness of the composite maintained nearly the same level. The main benefit of incorporating particulate titanium appeared to be the significant reduction in the slow crack growth rate. This can be important as the lifetime of an implant may be largely determined by the rate of slow crack growth.

1.5.4 Bioactive Metal Matrix Composites

Due to their excellent mechanical properties, corrosion resistance, and biocompability, titanium and titanium alloys have been used increasingly for dental and orthopedic implants. Titanium matrix composites with bioactive reinforcement were thus fabricated using a powder metallurgy route (Adoba et al., 1996, 1997). Bioactive phases in the composites included HA, Bioglass®, and Apoceram. The objective of producing these composites was to obtain bioactivity from the reinforcement and superior toughness from the metal matrix. Furthermore, designing bioactive functionally graded materials (FGMs) was investigated. Early work on a bioactive FGM was undertaken on a HA-Ti composite (Bishop et al., 1993). Fabrication of the FGM utilized cold pressing of layered mixtures of titanium and HA with a step-wise variation in HA concentration, followed by hot pressing. Improved mechanical properties were reported for these FGMs over the conventional composites. However, interfacial weakness between the layers was a problem. This problem may be partially solved by having a continuous variation in composition introduced via a vibration process (Adoba et al., 1996). Cold isostatic pressing and hot compaction could be used. The composites and FGMs fabricated at a relatively low temperature (500℃ and 750℃) were found to have a sintered Ti matrix with some porosity and unsintered HA agglomerates. The composite containing 10 wt% of HA fabricated at 1000℃ contained fewer pores in the Ti matrix, and the HA started to sinter. Despite the well-known reactivity of Ti, there was little evidence to suggest the presence of reaction products with HA. The starting powder of titanium may have a thin layer of oxide which served as a passivating layer and prevened reaction between HA and Ti. The failure of composites and FGMs after three-point bending tests was predominantly brittle fracture. The addition of HA into Ti decreased the bend strength of the matrix. FGMs demonstrated graceful failure with some load-carrying capacity after failure commenced. The fracture toughness values showed a significant improvement for the FGMs over the composites. Theoretical analysis showed that in FGMs the high HA content surface layers were under compressive residual stress. This accounted for the higher fracture toughness values of the FGMs compared to their conventional composite counterparts. Furthermore, the interface between the layers in an FGM hindered primary crack propagation. For bioactivity testing, samples of HA-Ti, Apoceram-Ti, and Bioglass®-Ti were immersed in Ringer's solution and subsequently analysed using FTIR and FT-Raman. The spectra showed a phosphate peak from the surface carbonate apatite layer in composites containing high reinforcement contents (i. e. 20%) after 24 hours of immersion. The peak intensity increased with immersion time and became well defined after 120 hours of immersion.

1.6 Concluding Remarks

Materials of all categories have been used in the medical field. At present, bioactive materials are mainly bioactive ceramics, glasses, and glass-ceramics, which are specially designed to interact with body tissue after their implantation. These materials can take the form of monolith (dense or porous), particulate, coating, and the matrix or reinforcement of a composite. In all circumstances, the biocompatibility and bioactivity of these materials as well as their derivatives are of prime importance, as implants made of these materials are expected to elicit favorable response from the surrounding tissue and establish a stable (and chemical) bond with the tissue within a relatively short period. As has been shown in this chapter, various bioactive materials have their respective advantages and disadvantages as far as their properties (i. e. biological, mechanical and other properties) and applications are concerned. Sometimes, a compromise has to be made in order to use a specific material to solve a particularly difficult medical problem.

Apart from the materials discussed in this chapter, there are three other (types of) materials that require consideration when the bioactivity or potential bioactivity of a material is explored: (a) synthetic polymers such as Polyactive®, (b) titanium, and (c) surface-treated metal such as tantalum. Polyactive® is a block polymer of poly(ethylene oxide) and poly(butylene teraphthalate) and is bone-bonding. Other polymeric materials such as silicone also show signs of bioactivity after some chemical treatment. Osteointegration of titanium was discovered more than 20 years ago. and the use of this concept has extended from dentistry to other medical fields where a direct bone-to-biomaterial interface (without fibrous tissue) is observed. After appropriate surface (and heat) treatment, certain metals such as tantalum can induce apatite layer formation *in vitro* and hence are expected to form chemical bonds with tissue after implantation. This recent development is significant as metals are normally treated as nonbioactive materials.

As can be seen, bioactive materials, just like materials in general and biomaterials in particular, may eventually encompass materials in all categories that are known to humankind: ceramics, polymers, metals, and composites. Processing of bioactive materials may not require special, advanced technology. But extreme care needs to be taken in producing, handling and storing these materials. Material synthesis, chemical analysis, mechanical evaluation, biological testing, long-term property assessment, etc. are all integral parts of the effort in developing new bioactive materials.

Acknowledgements

The author is grateful to Professor William Bonfield for leading him to the biomaterials field and for exposing him to various aspects of biomaterials R & D during his time as a researcher in one of the leading biomaterials centres in the world. He sincerely thanks his colleagues and students in the University of London and Nanyang Technological University for their assistance and discussions. Funding for his research in bioceramics, bioactive coatings and bioactive composites from various sources is gratefully acknowledged. He is indebted to his wife, Dr. Ming Li, for her encouragement, support, care, and patience.

References

Abram, J., J. Bowman, J. C. Behiri, W. Bonfield. Plast. Rubber Proc. Appl. 4: 261 – 269 (1984)

Adoba, A. E., H. B. McShane, R. D. Rawlings, I. U. Rehman. *Proc. 11th Int. Conf. Composite Mater.* Gold Coast, Australia, 1:496 – 508 (1997)

Adoba, A. E., H. B. McShane, R. D. Rawlings. *Proc*, *7th Eur*. Conf. *Composite Mater*. London, UK, 1: 461 – 466 (1996)

Andersson, Ö. H., PhD Dissertation. Âbo Akademi. Finland (1990)

Anthony, P. P., G. A. Gie, C. R. Howie, R. S. M. Ling. J. Bone. Surg. 72-B:971 – 979 (1990)

Aoki, H.. *Medical Applications of Hydroxyapatite.* Ishiyaku EuroAmerica, Tokyo, St. Louis (1994)

Arita, I. H., V. M. Castano, D. S. Wilkinson. J. Mater. Sci. Mater. Med. 6:19 – 23 (1995)

Bajpai, P. K., W. G. Billotte. In *The Biomedical Engineering Handbook*. Ed. by J. D. Bronzino. CRC Press, Boca Raton, Fla. 552 – 581 (1995)

Behiri, J., M. Braden, S. N. Khorasani, D. Wiwattandate, W. Bonfield.

Bioceramics. 4:301 –304 (1991)
Berndt, C. C., K. A. Gross. Proc. Int. Thermal Spray Conf. Exposition. USA, pp. 465 –470 (1992)
Best, S., B. Sim, M. Kayser, S. Downes. J. Mater. Sci. Mater. Med. 8:97 – 103 (1997)
Best, S., W. Bonfield. J. Mater. Sci. Mater. Med. 5:516 –512 (1994)
Bishop, A. M., C. Y. Lin, M. Navaratnam, R. D. Rawlings, H. B. McShane. J. Mater. Sci. Lett. 12:1516 – 1518 (1993)
Black, J.. *Biological Performance of Materials*. 2nd ed., Marcel Dekker, New York (1992)
Black, J., G. Hastings. eds. *Handbook of Biomaterial Properties*. Chapman & Hall, London (1998)
Bohner, M., J. Lemaître. J. Mater. Sci. Mater. Med. 7:457 –463 (1996)
Bonfield, W., C. Doyle, K. E. Tanner. In: P. Christel, A. Meunier and A. J. C. Lee, eds. *Biological and Biomechanical Performance of Biomaterials*. Elsevier Science, Amsterdam, pp. 153 – 158 (1986)
Bonfield, W., M. D. Grynpas, A. E. Tully, J. Bowman, J. Abram. Biomaterials. 2: 185 – 186 (1981)
Bonfield, W., M. D. Grynpas, J. A. Bowman. U. K. Pat. GB2085461B (1984)
Boyne, P. J., B. A. Shapton. Trans. 4th Annu. Meet. Soc. Biomat. (1978)
Brezny, R., D. J. Green. J. Am. Ceram. Soc.. 72:1145 (1989)
Bromer, H., E. Pfeil, H. H. Kas. German Pat. 2326100 (1973)
Brossa, F., A. Cigada, R. Chiesa, L. Paracchini, C. Consonni. *Bio-Medical Materials and Engineering*. 3:127 – 136 (1993)
Brown, S. R., I. G. Turner, H. Reiter. J. Mater. Sci. Mater. Med. 5:756 – 759 (1994)
Bucholz, R. W., A. Charlton, R. E. Holmes. Orthoped. Clin. N. Am. 182:323 – 334 (1987)
Bunker, B. C., D. R. Tallant, T. J. Headley, G. L. Turner, R. J. Kukpatrik. Phys. Chem. Glasses. 29:106 – 120 (1988)
Capaccio, G., I. M. Ward. Polymer. 15:233 –238 (1974)
Carpenter, P. R., M. Campbell, R. D. Rawlings, P. S. Rogers. J. Mater. Sci. Lett. 21:1309 – 1312 (1986)
Castaldini, A., A. Cavallini. In: P. Christel, A. Meunier and A. J. C. Lee, eds. *Biological and Biomedical Performance of Biomaterials*. Elsevier Science. Amsterdam, Netherlands, pp. 525 –530 (1986)
Champion, E., S. Gautier, D. B. Assollant. J. Mater. Sci. Mater. Med. 7: 125 (1996)
Cheang, P., K. A. Khor. J. Mater. Process. Tech. 48:429 –436 (1995)
Chiroff, R. T., R. A. White, E. W. White, J. N. Weber, D. M. Roy. J. Biomed. Mater. Res. 11:165 – 178 (1977)
Clark, A. E., C. G. Pantano, L. L. Hench. J. Am. Ceram. Soc. 59:37 –39 (1976)
Clark, A. E., C. Y. Kim, J. K. West, J. Wilson, L. L. Hench. In: T. Yamamuro, J. Wilson and L. L. Hench, eds. *Handbook of Bioactive Ceramics*. Vol Ⅰ. CRC

Press, Boca Raton, FL. 73 (1990)
Clarke, K. I., S. E. Graves, A. T. C. Wong, J. T. Triffitt, M. J. O. Francis, J. T. Czernuszka. J. Mater. Sci. Mater. Med. 4:107 - 110 (1993)
Claxton, E., R. D. Rawlings, P. S. Rogers. *Proc. 7th Eur. Conf. Composite Mater.* London, UK, pp. 443 - 448 (1996)
Cook, S. D.. Clin. Orthoped. Relat. Res. 230:303 - 312 (1988)
Correia, R. N., M. C. F. Magalhaes, P. A. A. P. Marques, A. M. R. Senos. J. Mater. Sci. Mater. Med. 7:501 (1996)
Daculsi, G., E. Goyenvalle, E. Aguado. Bioceramics. 12:287 - 290 (1999)
Daculsi, G., N. Passuti, S. Martin, C. Deudon. J. Biomed. Mater. Res. 24:379 - 396 (1990)
Daculsi, G., R. Z. LeGeros, E. Nery, K. Lynch. J. Biomed. Mater. Res. 23:883 - 894 (1989)
de Groot, K.. Biomaterials. 1:47 (1980)
de Groot, K., C. P. A. T. Klein, J. G. C. Wolke, J. M. A. de Blieck-Hogervorst. In: T. Yamamuro, L. L. Hench and J. Wilson, eds. *Handbook of Bioactive Ceramics.* Vol Ⅱ. CRC Press, Boca Raton, FL. 133 - 142 (1990)
de Groot, K., R. Geesink, C. P. A. T. Klein, P. Serekian. J. Biomed. Mater. Res. 21:1375 - 1381 (1987)
de Lange, G. L., C. de Putter, E. H. Burger, K. de Groot. In: A. Pizzoferrato, et al., eds. Proc. *Sixth Eur. Conf. Biomaterials.* Elsevier Science, Amsterdam, pp. 217 - 221 (1987)
de With, G.. J. Mater. Sci. 16:1592 (1981)
Deer, W. A., R. Ahowie, J. Zussman. *An Introduction to the Rock Forming Minerals.* Longman, Hongkong, pp. 504 - 509 (1985)
Delecrin, J., D. Daculsi, J. Pouezat, N. Passuti, B. Duquet, S. Szmuckler-Moncler. Bioceramics. 4:317 - 323 (1991)
Denissen, H. W., K. de Groot. In *Mechanical Properties of Biomaterials.* John Wiley, Chichester, New York (1980c)
Denissen, H. W., H. J. A. Van Kijk, K. de Groot, P. J. Klopper. In: *Mechanical Properties of Biomaterials.* John Wiley, Chichester, New York, pp. 489 - 505 (1980b)
Denissen, H. W., K. de Groot. *The 1st World Biomater. Congress.* Paper 3. 8. 1 (1980a)
Dove, J. H., S. F. Hulbert. *Proc. 2nd Int. Symp. Apatite.* Tokyo, Japan (1995)
Downes, R. N., S. Vardy, K. E. Tanner, W. Bonfield. Bioceramics. 4:239 - 246 (1991)
Downes, S., D. J. Wood, A. J. Malcolm, S. Y. Ali. Clin. Orthoped. Rel. Res. 252:294 - 298 (1990)
Doyle, C., K. E. Tanner, W. Bonfield. Biomaterials. 12:841 - 847 (1991)
Ducheyne, P.. J. Biomed. Mater. Res: Appl. Biomater. 21:219 - 236 (1987)
Ducheyne, P., L. L. Hench, A. Kagan, M. Martens, A. Bursens, J. C. Mulier. J. Biomed. Mater. Res. 14:225 (1980)
Eggli, P. S., W. Muller, R. K. Schink. Clin. Orthop. Rel. Res. 232: 127 - 138 (1988)
Elliott, J. C.. *Structure and Chemistry of the Apatites and Other Calcium Crthophosphates.*

Elsevier Science, Amsterdam (1994)
Fabbri, M., G. C. Celotti, A. Ravaglioli. Biomaterials. 16:225 (1995)
Fang, F., D. K. Agrawal, D. M. Roy. In: P. W. Brown and B. Constantz, eds. *Hydroxyapatite and Related Materials*. CRC Press, Boca Raton, FL, pp. 269 – 282 (1994)
Fawcett, D. W. *A Textbook of Histology*. W. B. Company, London (1986)
Fowler, B. O. Inorg. Chem. 13:194 (1974)
Frank, F. C. Proc. R. Soc. Lond. A, 319:127 – 136 (1970)
Friedrich, K., U. A. Karsch. J. Mater. Sci. 16:2167 – 2175 (1981)
Fu, Q., G. Wang. Polym. Eng. Sci. 32:94 – 97 (1992)
Furusawa, T, K. Mizunuma. Implant Dent. 6:93 – 101 (1997)
Gatti, A. M., D. Zaffi, C. P. Poli. Biomaterials. 11:513 (1990)
Geesink, R. G. T.. Clin. Orthoped. 261:39 – 58 (1991)
German, R. M. *Powder Injection Molding*. Metal Powder Industries Federation, Princeton, NJ (1990)
Gibson, A. G., I. M. Ward, B. N. Cole, B. Parsons. J. Mater. Sci. Lett. 9:1193 – 1196 (1974)
Gibson, I. R., M. Akao, S. M. Best, W. Bonfield. Bioceramics. 9:173 (1996)
Goodwin, C. J., M. Braden, S. Downes, N. J. Marshall. J. Biomed. Mater. Res. 34:47 – 55 (1997)
Greco, R. S., ed.. Implantation biology. CRC Press, Boca Raton, FL. (1994)
Greenspan, D. C., L. L. Hench,. J. Biomed. Maters. Res.. 10:503 – 509 (1976)
Gross, U., J. Brandes, V. Strunz, I. Bab, J. Sela. J. Biomed. Mater. Res. 15:291 – 305 (1981)
Gross, U., V. Strunz. J. Biomed. Mater. Res. 19:251 – 271 (1985)
Gross, U. M., C. Muller-Mai, C. Voigt. In: L. L. Hench and J. Wilson, eds. *An Introduction to Bioceramics*. World Scientific, Singapore (1993)
Gross, U. M., V. Strunz. J. Biomed. Mater. Res. 14:607 – 618 (1980)
Grote, J. J., W. Kuijpers, K. de Groot. Q. R. S. 43:248 (1981)
Gruen, T. A., G. M. MeNeice, H. C. Amstutz. Clin. Orthoped. Relat. Res. 141:17 – 27 (1979)
Guild, F. J., R. J. Young. J. Mater. Sci. 24:298 – 306 (1989a)
Guild, F. J., R. J. Young. J. Mater. Sci. 24:2454 – 2460 (1989b)
Guild, F. J., W. Bonfield. Biomaterials. 14:985 – 993 (1993)
Guillemin, G. F., L. Patat, J. Fournie, M. Chetail. J. Biomed. Mater. Res. 21:557 – 567 (1987)
Helmus, M. N., K. Tweden. In: D. L. Wise, ed. *Encyclopedic Handbook of Biomaterials and Bioengineering*. Marcel Dekker, New York, 1:27 – 59 (1995)
Hench, L. L. H. A. Paschall. *J. Biomed. Mater. Res. Symp.* 4:25 – 42 (1973)
Hench, L. L., A. E. Clark. In: *Biocompatibility of Orthopedic Implants*, Vol 2. D. F. Williams. CRC Press, Boca Raton, FL (1982)
Hench, L. L., D. E. Clark. J. Non-Cryst. Solids. 28:83 (1978)
Hench, L. L., E. C. Ethridge. *Biomaterials: An Interfacial Approach*. Academic Press, New York (1982)
Hench, L. L.. In: B. D. Ratner, A. S. Hoffman, F. J. Schoen and J. E. Lemons, eds. *Biomaterials Science: An Introduction to Materials in Medicine*. Academic

Press, San Diego (1996)
Hench, L. L.. J. Am. Ceram. Soc. 74:1487 – 1510 (1991)
Hench, L. L., J. Wilson, eds. *An Introduction to Bioceramics.* World Scientific, Singapore (1993)
Hench, L. L., R. J. Splinter, W. C. Allen, T. K. Greenlee, Jr. J. Biomed. Mater. Res. 2:117 – 141 (1972)
Henrich, D. E., A. E. Cram, J. B. Park, Y. K. Lui, H. Reddi. J. Biomed. Mater. Res.. 27:277 – 280 (1993)
Herman, H.. Sci. Am. 259:112 – 117 (1988)
Hing, K., S. M. Best, K. E. Tanner, P. A. Revell, W. Bonfield. Bioceramics. 9: 157 – 160 (1996)
Hing, K. A., S. M. Best, K. E. Tanner, P. A. Revell, W. Bonfield. Bioceramics. 8: 75 – 80 (1995)
Holand, W., P. Wange, K. Naumann, J. Vogel, G. Carl, C. Jana, W. Gotz. J. Non-Crystalline Solids. 129:152 – 262 (1991)
Holand, W., W. Vogel. In: L. L. Hench and J. Wilson, eds. *An Introduction to Bioceramics.* World Scientific, Singapore, pp. 125 – 137 (1993)
Holmes, R. E. Plast. Reconstr. Surg. 63:626 – 633 (1979)
Hoogendoorn, H. A., W. J. Visser, S. J. Oldenburg, L. N. A. Akkermans, P. Wittebol, W. Renooij. In: A. Pizzoferrato et al., eds. *Proc. Sixth Eur. Conf. Biomater.* Elsevier Science, Amsterdam, pp. 81 – 86 (1987)
Huang, J., L. Di Silvio, M. Wang, K. E. Tanner, W. Bonfield. J. Mater. Sci. Mater. Med. 8:775 – 779 (1997)
Huaxia, J.. J. Mater. Sci. Mater. Med. 3:283 – 287 (1992)
Hubbard, W. *Physiological Calcium Phosphates as Orthopedic Biomaterials*, PhD Thesis, Marquette University, Milwallkee, WI, USA (1974)
Huiskes, R., H. Weinans, B. van Rietbergen. Clin. Orthopaed. Rel. Res. 274:124 (1992)
Hulbert, S. F., J. C. Bokros, L. L. Hench, J. Wilson, G. Heimke. In *High Tech Ceramics.* Elsevier Science, Amsterdam, pp. 3 – 27 (1987)
Inoue, O., H. Shimabukuro, Y. Shingaki, K. Ibaraki. Bioceramics. 5:411 – 418 (1992)
Inoue, O., K. Ibaraki, H. Shimabukuro, Y. Shigaki. Bioceramics. 4:247 – 254 (1991)
IoM Report. *Materials Technology Foresight in Biomaterials.* Institute of Materials, London, UK (1995)
Ito, A., K. Teraoka, S. Tsutsumi. Bioceramics. 9:189 – 192 (1996)
James, P. J.. ed. *Isostatic Pressing Technology.* Applied Science, London (1983)
Janicke, S., W. Wagner. Adv. Biomater. 8:67 – 69 (1988)
Jarcho, M., C. H. Bolen, M. B. Thomas, J. Bobick. J. Mater. Sci. 11:20 – 27 (1976)
Jarcho, M., J. F. Kay, K. I. Gumaer, R. H. Doremus. J. Bioeng. 1:79 (1977)
Jiang, G., D. Shi. J. Biomed. Mater. Res. Appl. Biomater. 43:77 – 81 (1998)
Kasperk, D., R. Ewers. *Proc. Sixth Eur. Conf. Biomaterials.* ed. by A. Pizzoferrato et al. Elsevier Science, Amsterdam, 197 – 203 (1987)
Kawanabe, K., J. Tamura, T. Yamamuro, T. Nakamura, T. Kokubo, S. Yoshihara. J. Appl. Biomater. 4:135 – 141 (1993)

Kay, J. F. Handbook of Bioactive Ceramics. Vol. Ⅱ, ed. by T. Yamamuro, L. L. Hench and J. Wilson, (CRC Press, Boca Raton, FL 1990),111 –122
Kent, J., R. James, I. Finger, M. Jarcho. The 1st World Biomaterial Congress Paper 3.8.2 (1980)
Khor, K. A., P. Cheang. J. Thermal Spray Technol. 3:45 –50 (1994)
Khor, K. A., P. Cheang, N. Wang. Journal of Thermal Spray Technol. 7:254 –260 (1997)
Khor, K. A., P. Cheang. *Proc. 7th Nat. Thermal Spray Conf.* USA, 153 – 157 (1994)
Khor, K. A., P. Cheang, Y. Wang. J. Minerals, Metals & Materials Society. 49:51 (1997)
Khorasani, S. N., S. Deb, J. C. Behiri, M. Braden, W. Bonfield. Bioceramics. 5: 225 –232 (1992)
Kitsugi, T., T. Yamamuro, T. Nakamura, S. Higashi, T. Kakutani, Y. Hyakuna, S. Ito, T. Kokubo, T. Shibuya. J. Biomed. Mater. Res. 21:467 –484 (1987a)
Kitsugi, T., T. Nakamura, T. Yamamuro, T. Kokubo, T. Shibuya, M. Takagi. J. Biomed. Mater. Res. 21:1255 –1271 (1987b)
Kitsugi, T., T. Yamamuro, T. Nakamura, S. Higashi, Y. Kakutani, K. Hyakuna, S. Ito, T. Kokubo, M. Takagi, T. Shibuya. J. Biomed. Mater. Res. 20:1295 –1307 (1986)
Klein, C. P. A. T., P. Patka, W. den Hollander. Biomaterials. 10:59 –62 (1989)
Knabe, C., W. Ostapowicz, R. J. Radlanski. J. Mater. Sci. Mater. Med.. 9:337 –345 (1998)
Knowles, J. C., W. Bonfield. J. Biomed. Mater. Res. 27:1591 –1598 (1993)
Koch, B., J. G. C. Wolke, K. de Groot. J. Biomed. Mater. Res. 24:655 – 667 (1990)
Kokubo, T.. Biomaterials. 12:155 –163 (1991)
Kokubo, T., C. Ohtsuki, S. Kotani, T. Kitsugi, T. Yamamuro. *Bioceramics*. ed. by G. Heimke, German Ceramic Society, 2:113 –120 (1990)
Kokubo, T., H. Kushitani, C. Ohtsuki, S. Sakka, T. Yamamuro. J. Mater. Sci. Mater Med. 3:95 –100 (1992)
Kokubo, T.. In: L. L. Hench and J. Wilson, eds. *An Introduction to Bioceramics*. World Scientific, Singapore (1993)
Kokubo, T. J. Non-Cryst. Solids. 120:138 –151 (1990)
Kokubo, T., M. Shigematsu, Y. Nagashima, M. Tashiro, T. Nakamura, T. Yamamuro, S. Higashi. Bull. Inst. Chem. Res. Kyoto Univ. 60:260 – 268 (1982)
Kokubo, T., S. Ito, M. Shigematsu, S. Sakka, T. Yamamuro. J. Mater. Sci. 22: 4067 –4070 (1987)
Kokubo, T., S. Ito, M. Shigematsu, S. Sakka, T. Yamamuro. J. Meter. Sci. 20: 2001 –2004 (1985)
Kokubo, T., S. Ito, S. Sakka, T. Yamamuro. J. Meter. Sci. 21:536 –540 (1986)
Kumar, R. R., M. Wang. *Trans. 6th World Biomater. Congr*. Hawaii, USA, pp. 1310 (2000a)
Kumar, R. R., M. Wang. Bioceramics. 13:(2000b)
Kumar, R. R., M. Wang, P. Ducheyne. Bioceramics. 13:(2000c)

Lacefield, W. R.. In: L. L. Hench and J. Wilson, eds. *An Introduction to Bioceramics*. World Scientific, Singapore, pp. 223 – 238 (1993)
Ladizesky, N. H., M. Wang, I. M. Ward, W. Bonfield. Med. Biol. Eng. Comput. 34: Suppl. 1, part 2, 155 – 156 (1996)
Lange, F. F., K. T. Miller. Adv. Ceram. Mater. 2:827 (1987)
LeGeros, R. Z.. Calcium phosphates in oral biology and medicine. In: H. Myers, ed. *Monographs in Oral Sciences*, Vol. 15. S. Karger, Basel (1991)
LeGeros, R. Z. Proc. Int. Semi. Orthopaed. Res. Nagoya (1990). ed. S. Niwa. Springer-Verlag (1992)
Lemaître, J., M. Maréchalle. Bioceramics. 9:197 (1996)
Lin, C. K., C. C. Berndt. J. Thermal Spray Technol. 3:75 – 104 (1994)
Lin, J. H. C., M. L. Liu, C. P. Ju. J. Mater. Sci. Mater. Med. 5:279 – 283 (1994)
Liu, D. M., H. M. Chou, J. D. Wu. J. Mater. Sci. Mater. Med. 5:147 – 153 (1994)
Liu, D. M. In: *Porous Ceramic Materials: Fabrication, Characterization, Applications*. Trans Tech Publications Ltd., Aedermannsdorf, pp. 209 – 232 (1996)
Liu, D. M., J. J. Brown. Mater. Chem. Phys. 32:161 (1992)
Locardi, B, U. E. Pazzaglia. Biomaterials. 14:437 – 441 (1993)
Luo, L., T. G. Nieh. Mater. Sci. Eng. C, 3:77 – 78 (1995)
Maiti, S. N., K. K. Sharma. J. Mater. Sci. 27:4605 – 4613 (1992)
Maxian, S. H., J. P. Zawadsky. J. Biomed. Mater. Res. 27:717 – 728 (1993)
Mendelon, M. I., T. N. McKechnie, L. B. Spiegel. *Cerami. Eng. Sci. Proc.* 15: (5), 555 (1994)
Merwin, G. E. In: T. Yamamuro, L. L. Hench and J. Wilson, eds. *Handbook of Bioactive Ceramics*. Vol I. CRC Press, Boca Raton, FL, pp. 323 – 328 (1990)
Mittal, K. L.. Electrocomp. Sci. Technol. 3:21 (1976)
Monma, H.. J. Ceram. Soc. Jpn: Int. Ed., 101:718 – 720 (1993)
More, E. A., W. Votava, D. B. Bass, J. McMullen. J. Dent. Res. 50:260 (1971)
Muller-Mai, C., D. Amir, H. Wendland, Z. Schwartz, S. Sela, U. Gross. J. Biomed. Mater. Res. 24:1571 – 1584 (1990)
Muller-Mai, C., D. Amir, Z. Schwartz, J. Sela, B. D. Boyan, H. Wendland, U. M. Gross. Cells Mater. 1:341 – 352 (1991)
Muller-Mai, C., H. J. Schmitz, V. Strunz, G. Fuhrmann, T. Fritz, U. M. Gross. J. Biomed. Mater. Res. 23:1149 – 1168 (1989)
Murakami, A., J. Behiri, W. Bonfield. In: S. A. Paipetis and G. C. Papanicolaou, eds. *Phase Interaction in Composite Material*. Omega Scientific, Oxford, pp. 460 – 466 (1992)
Nakamura, T., T. Yamamuro, S. Higashi, T. Kokubo, S. Ito. J. Biomed. Mater. Res. 19:685 – 698 (1985)
Nakashima, Y.. J. Biomed. Mater. Res. 35:287 – 298 (1997)
Neo, M., S. Kotani, Y. Fujita, T. Nakamura, T. Yamamuro, Y. Bando, C. Ohtsuki, T. Kokubo. J. Biomed. Mater. Res.. 26:255 – 267 (1992)
Nerf, E. B., K. L. Lynch, W. M. Hirthe, K. H. Muller. J. Periodontal. 46:328 (1975)
Ogino, M., F. Ohuchi, L. L. Hench. J. Biomed. Maters. Res. 14:55 – 64 (1980)
Ogiso, M., H. Kaneda, J. Arasaki. Trans. Soc. Biomat. 4:54 (1981)
Oguchi, H. Biomaterials. 13:471 – 477 (1992)

Ohtsuki, C., T. Kokubo, T. Yamamuro. J. Non-Cryst. Solids. 143:84 – 92 (1992)
Ohura, K., T. Nakamura, T. Yamamuro, T. Kokubo, Y. Ebisawa, Y. Kotoura, M. Oka. J. Biomed. Mater. Res. 25:357 – 365 (1991)
Olmi, R., A. Moroni, A. Castaldini, R. Romagnoli. In: P. Vincenzini, ed. *Ceramics in Surgery*. Elsevier Science, Amsterdam, Netherlands, pp. 91 – 96 (1983)
Ong, J. L., L. C. Lucas, W. R. Lacefield, E. D. Rigney. Biomaterials. 13:249 – 254 (1992)
Oonishi, H.. In: J. E. Davies, ed. *The Bone-Biomaterial Interface*. University of Toronto Press, Toronto, Canada, pp. 321 – 333 (1991)
Park, H. C., Y. K. Liu, R. S. Lakes. J. Biomech. Eng. 108:141 – 148 (1986)
Park, J. B., R. S. Lakes. *An Introduction to Biomaterials*, 2nd ed. Plenum Press, New York (1992)
Pawlowski, L. The Science and Engineering of Thermal Spray Coatings. John Wiley & Sons, New York (1995)
Piotrowski, G., L. L. Hench, W. C. Allen, G. J. Miller. J. Biomed. Mater. Res. 9:47 – 61 (1975)
Ratner, B. D, A. S. Hoffman, F. J. Schoen, J. E. Lemons, eds. *Biomaterials Science: An Introduction to Materials in Medicine*. Academic Press, San Diego (1996)
Ravaglioli, A., A. Kraejewski. *Bioceramic: Materials, Properties and Applications*. Chapman & Hall, London (1992)
Reck, R. Laryngoscope. 94:(2), Suppl. 33, 1 (1984)
Reck, R., S. Storkel, A. Meyer. Ann. N. Y. Acad. Sci. 523:100 (1988)
Rejda, B. V., J. G. J. Peelen, K. de Groot. J. Bioeng. 1:93 (1977)
Revell, P. A., K. E. Tanner. In: A. Ravaglioli and A. Krajewski, eds. *Cells, Ceramics and Tissues*. (1998)
Rickerby, D. S.. Surf. Coat. Technol. 36:541 – 557 (1988)
Riley, F. L.. In: *Progress in Nitrogen Ceramics*. NATO ASI Series, Boston, Martinus Nijhoff, pp. 121 – 133 (1983)
Rivero, D. P., J. Fox. J. Biomed. Mater. Res. 22:191 – 202 (1988)
Roberts, G. A. F. *Chitin Chemistry*. Macmillan, London (1992)
Roy, D. M., S. A. Linnehan. Nature. 247:220 – 227 (1974)
Royer, P., S. Amrah-Bouali, M. Freche, C. Rey, N. Rouquet, G. Bonel. Bioceramics. 5:95 – 102 (1992)
Salomoni, A., A. Tucci, L. Esposito, I. Stamenkovic. J. Mater. Sci. Mater. Med. 5:651 (1994)
Santos, J. D., J. C. Knowles, S. Morrey, F. J. Monterio, G. W. Hastings. Bioceramics. 5:35 – 41 (1992)
Santos, J. D., L. J. Jha, F. J. Monteiro. J. Mater. Sci. Mater. Med. 7:181 – 185 (1996)
Schepers, E., M. DeClerco, P. Ducheyne, R. Kempeneers. J. Oral Rehabil. 18: 439 – 452 (1991)
Schepers, E. J. G., P. Ducheyne. Bioceramics. 6:401 – 404 (1993)
Schepers, E. J. G., P. Pinruethal. Bioceramics. 6:113 – 116 (1993)
Shankar, S., D. E. Koenig, L. E. Dardi. J. Met. 13 (1981)
Shaw, J., S. M. Best, W. Bonfield, A. Marsh, J. Cotton. J. Mater. Sci. Lett. 14: 1055 – 1057 (1995)

Shepers, E. J. G., P. Ducheyne. J. Oral Rehabil. 24:171 – 181 (1997)
Sherman, A. J., R. H. Tuffias, R. B. Kaplan. Am. Ceram. Soc. Bull. 69: 1025 (1991)
Shi, D., G. Jiang, X. Wen. *Proc. 8th Int. Conf. Process. Fabrication Adv. Mater.* Singapore (1999)
Shinjo, K., T. Asai, S. Saito, M. Tukamoto. Bioceramics. 4:263 (1991)
Silva, V. V., R. Z. Domingues. J. Mater. Sci. Mater. Med. 8:907 (1997)
Singh, R. K., F. Qian, V. Nagabushnam, R. Damodaran, B. Moudgil. Biomaterials. 15:522 – 528 (1994)
Slosarczyk, A., J. Bialoskorski. J. Mater. Sci. Mater. Med. 9:103 (1998)
Sogal, A., S. F. Hulbert. Bioceramics. 5:213 – 224 (1992)
Stanley, H. R., M. B. Hall, F. Colaizzi, A. E. Clark. J. Prosthetic Dentistry. 58:607 – 613 (1987)
Suga, Y., T. Harjanto, J. Takahashi. *Thermal Spray: International Advances in Coating Technology*. C. C. Berndt, ed. ASM International, USA, pp. 247 – 252 (1992)
Suwanprateeb, J., K. E. Tanner, S. Turner, W. Bonfield. J. Mater. Sci. Mater. Med. 6:804 (1995)
Suwanprateeb, J., K. E. Tanner, S. Turner, W. Bonfield. J. Mater. Sci. Mater. Med. 8:469 (1997)
Swain, R. E., M. Wang, B. Beale, W. Bonfield. Biomed. Engi.: Appli. Basis Commun. 11:315 – 320 (1999)
Tagai, H., H. Aoki. In: *Bioceramics Symp.*, Vol. 16. University of Keele, UK, (1978)
Tampieri A., G. Celotti, F. Szontagh, E. Landi. J. Mater. Sci. Mater. Med.. 8:29 – 37 (1997)
Tang, F., B. Chua, M. Wang. *Trans. 6th World Biomater. Congr.* Hawaii, p. 539 (2000)
Tanner, K. E., D. Vashishth, P. T. Ton That, W. Bonfield. *Trans. 6th World Biomater. Congr.* Hawaii, p. 408 (2000)
Tanner, K. E., R. N. Downes, W. Bonfield. Br. Ceram. Trans. 93: 104 – 107 (1994)
Taylor, B. A. PhD thesis. Imperial College. London (1994)
Thomas, K. A.. J. Biomed. Mater. Res. 21:1395 – 1414 (1987)
Thomas, M. B., R. H. Doremus, M. Jarcho. J. Mater. Sci. 15:891 (1980)
Toth, J. M., K. L. Lynch, T. R. Devine. In: D. L. Wise, ed. *Encyclopedic Handbook of Biomaterials and Bioengineering*. Marcel Dekker, New York, Vol. 2, Part A, 1466 – 1501 (1995)
Tracey, M. V. Rev. Pure Appl. Chem. 7:1 – 15 (1957)
Tucker, R. C.. In: R. F. Bunshah, ed. *Handbook of Deposition Technologies for Films and Coatings*. Noyes, Park Ridge, pp. 591 – 642 (1994)
Uenu, Y., Y. Shima, T. Akiyama. In: *High Tech Ceramics*. Elsevier Science, Amsterdam, pp. 369 – 379 (1987)
UK OST Report. Technology foresight-materials. Office of Science and Technology, HMSO (1995)
Van Landuyt, P., F. Li, J. P. Keustermans, J. M. Streydio, F. Delannay. J. Mater.

Sci. Mater. Med. 6:8 – 13 (1995)
Van Landuyt, P., G. Pouchkine, J. Lemaitre. Bioceramics. 11:247 – 250 (1998)
Vogel, J., C. Russel, G. Gunther, P. Hartmann, F. Vizethum, N. Bergner. J. Mater. Sci. Mater. Med. 7:495 – 499 (1996)
Vogel, W., W. Holand. Angew. Chem. Int. Ed. Engl. 26:527 – 544 (1987)
Wang, C. K., J. H. Lin, C. P. Ju, H. C. Ong, R. P. H. Chang. Biomaterials. 18: 1331 – 1338 (1997)
Wang, C. X., Z. Q. Chen, M. Wang. Trans. *6th World Biomateri. Congr.* Hawaii, pp. 1037 (2000a)
Wang, C. X., Z. Q. Chen, M. Wang. Bioceramics. 13:(2000b)
Wang, M., R. Joseph, W. Bonfield. Biomaterials. 19:2357 – 2366 (1998a)
Wang, M., C. Berry, M. Braden, W. Bonfield. J. Mater. Sci. Mater. Med. 9:621 – 624 (1998b)
Wang, M., C. X. Wang, J. Weng, J. Ni. *Trans. 6th World Biomater. Congr.* Hawaii, pp. 81 (2000)
Wang, M., C. Y. Yue, B. Chua, L. C. Kan. Bioceramics. 12:401 – 404 (1999)
Wang, M., D. Porter, W. Bonfield. Br. Ceram. Trans.. 93:91 – 95 (1994)
Wang, M., I. M. Ward, W. Bonfield. *Proc. 11th Int. Conf. Composite Mater.* Gold Coast, Australia, 1:488 – 495 (1997)
Wang, M., W. Bonfield, L. L. Hench. Bioceramics. 8:383 – 388 (1995)
Wang, M., S. Deb, K. E. Tanner, W. Bonfield. Proc. *7th Eur. Conf. Composite Mater.* London, 2:455 – 460 (1996a)
Wang, M., W. Bonfield, T. Kokubo. Bioceramics. 9: 387 – 390 (1996b)
Wang, M., X. Y. Yang, K. A. Khor, Y. Wang. J. Mater. Sci. Mater. Med. 10: 269 – 273 (1999)
Wen, H. B., J. R. de Wijin, F. Z. Cui, K. de Groot. Biomaterials. 19: 215 – 221 (1998)
Weng, J., M. Wang. Bioceramics. 13:(2000)
Weng, J., Q. Liu, J. G. C. Wolke, X. D. Zhang, K. de Groot. Biomaterials. 18: 1027 – 1035 (1997)
Williams, D. F. (ed.). Definitions in Biomaterials. Elsevier Science, Amsterdam, (1987a)
Williams, D. F.. Mater. Sci. Technol. 3:797 – 806 (1987b)
Williams, J. L., W. J. H. Johnson. J. Biomechanics. 22: 673 – 682 (1989)
Wilson, J., D. Nolletti. In: T. Yamamuro, L. L. Hench and J. Wilson, eds. *Handbook of Bioactive Ceramics.* CRC Press, Boca Raton, FL. (1990)
Wilson, J., G. E. Merwin. J. Appl. Biomater. 22A:159 – 177 (1988)
Wilson, J., S. B. Low. J. Appl. Biomater. 3:123 – 129 (1992)
Wilson, J., E. Douek, K. Rust. Bioceramics. 8:239 – 245 (1995)
Wolff, J.. *Das Gesetz der Transformation der Knochen.* Hirschwald, Berlin (1892); Reprinted in English Translation. Springer-Veriag (1987)
Wolke, J. G. C., C. P. A. T. Klein, K. de Groot. Proc. 3rd Natl. Thermal Spray Conf. USA, pp. 413 – 417 (1990)
Wolke, J. G. C., V. J. P. C. der Waerden, K. de Groot, J. A. Jansen. Biomaterials. 18: 483 – 488 (1997)
Xavier, S. F., J. M. Schultz, K. Friedrich. J. Mater. Sci. 25:2428 – 2432 (1990)

Yamamuro, T.. In: L. L. Hench and J. Wilson, eds. *An Introduction to Bioceramics*. World Scientific, Singapore (1993)
Yamamuro, T.. In: T. Yamamuro, J. Wilson and L. L. Hench, eds. *Handbook of Bioactive Ceramics*, Vol Ⅰ. CRC Press, Boca Raton, FL, 335 - 342 (1990)
Yang, H. Y., M. Wang. Bioceramics. 12:349 - 352 (1999b)
Yang, H. Y., M. Wang. Proc. *8th Int. Conf. Process. Fabrication Adv. Mater*. Singapore (1999a)
Zeiler, G., V. Strunz, U. Gross. In: R. Kossowsky and N. Kossowsky, eds. *Material Sciences and Implant Orthopedic Surgery*. NATO ASI Series, Martinus Nijhoff, Dordrecht, 249 - 259 (1986)
Zyman, Z., Y. Cao, X. Zhang. Biomaterials. 14:1140 - 1144 (1993)

2 Biocompatibility of Materials

Miqin Zhang

2.1 Introduction

2.1.1 Overview

Advances in biomaterials have been tremendous in both surgical and medical technologies over the past 30 years. Man-made materials and devices have been developed to replace parts of living systems in the human body, providing the patient the benefits of increased longevity and improved quality of life (Wise et al., 1995; Silver, 1994). Specific applications of biomaterials range from high-volume products such as blood bags, syringes, and needles to more challenging implantable devices that are used in cardiovascular, orthopedic, and dental applications as well as in a wide range of invasive treatments and diagnostic systems. An estimated 11 million people at a rate increasing by 5% – 15% each year in the United States have medical implants (Eisenberger, 1996; Wise et al., 1995). With such a tremendous increase in medical applications, the demand for a wide range of biomaterials is ever increasing, and the prospects for growth in this field is limitless. The categories of materials that are widely used as biomaterials include polymers, metals, ceramics, carbons, and their composites. As medical procedures become more sophisticated, the application of novel materials, such as prostheses or materials used in medical devices, are receiving increasing interest. The emergence of microelectronic and micro-electro-mechanical systems (MEMS) has stimulated a surge of interest in biomedical microdevices (BioMEMS). These microdevices offer opportunities for astounding medical advances: virus elimination, transplants of therapeutic cells, "magic bullets" that deliver cancer-fighting drugs on target, and implantable "biochips" for *in situ* diagnostics. BioMEMS hold a great promise for handling all aspects of engineering and biomedical problems. However, BioMEMS will reach their full potential only when the problems of their biocompatibility are completely solved (Bambot et al., 1995; Brody et al., 1995; Ferrari et al., 1996; Martin et al., 1999).

The host/biomaterial interactions which follow the implantation of any prosthesis or devices represent an important, complex, but comparatively poorly defined series of events. A clear understanding of the nature, complexity and significance of these interactions will lead to the development of improved biomaterials and better strategies for their evaluation and applications. The biocompatibility of materials is a complex topic, and much work has been devoted to this field during the past few decades. It is impossible to comprehensively review all the literature in this field. This chapter is therefore intended to give an overview of the biocompatibility of materials with emphasis on the important issues of biocompatibility for the materials commonly used in (BioMEMS) and implants. It is our hope that this chapter will provide the readers with information enough to understand the basic concepts of materials-biomolecular interactions and a framework for the researchers from various disciplines to further study the issues of biocompatibility to their specific applications. We will first introduce the concepts and important biomedical issues of biocompatibility. The host responses to materials and vice versa, materials responses to biological systems, will be discussed in detail to elucidate the nature and principles of protein adsorption, cell adhesion, blood – material and tissue – material interactions, as well as critical factors influencing material biocompatibility. This will provide a source of background information for the discussions of the topics in the subsequent sections and assist in understanding key characteristic features of the interactions between proteins, cells and materials. In this chapter, the commonly used methods for assessing biocompatibility are summarized. The performance of a foreign material in contact with biological fluids depends largely on protein adsorption. Hence, the issues related to protein adsorption at material – biological system interfaces are discussed with an emphasis on how the material surface properties influence the protein adsorption and how the adsorbed proteins affect subsequent cell adhesion. Biocompatible coatings commonly used to improve the biocompatibility of the materials are reviewed, with emphasis on the coatings for metal and silicon substrates, especially nonfouling PEG coatings used in implantable biomedical microdevices. Finally, a critical review of future trends in developing new surface modification protocol and methods of evaluating the biocompatibility of materials will be given.

2.1.2 Definitions and Concepts

Some important technical terms are defined in this subsection to facilitate further discussions on the issues of material biocompatibility. Their definitions have been either widely accepted or based on the author's own preference if more than one term is used for the same concept (Helmus and Tweden, 1995; Silver and Christiansen, 1999; Black, 1981).

1. Biomaterials

Biomaterials are any substances that can be used as a whole or as part of a system to treat, augment, or replace any tissue, organ, or function of the body with minimal adverse reaction or rejection by the body for an intended period of time.

2. Biomedical Devices or Implants

Biomedical devices or implants are devices that are engineered from biomaterials and designed to perform specific functions in the body.

3. Biological Performance

Biological performance refers to the host's and material's response: the local and systemic response of living systems to the material, and the response of the material to living systems, respectively.

4. Biostability

Biostability refers to the ability of a material to resist biodegradation mechanisms and maintain its properties *in situ*.

5. Bioinertibility

Bioinertibility refers to the ability of material not to cause harmful systematic toxic and carcinogenic effects on the host (e. g. corrosion and degradation), or to induce harmful response from the host (e. g. immune, inflammation, and allergic response). Bioinert materials require minimum non-specific protein adsorption and cell adhesion.

6. Biocompatibility

Biocompatibility is the ability of a material to perform with an appropriate host response in a specific application.

7. Bioactivity

Bioactivity refers to the inherent properties of a material to participate in specific biological reactions. Bioactive materials can promote the desired protein adsorption, cell attachment, and growth. For example, bioactive materials and coatings for orthopedic and dental implant applications consist of calcium phosphate ceramics. These materials promote biological fixation by dioectly binding with bones due to their chemical constituents similar to those of natural bone.

8. Biofouling

When the human body recognizes implanted materials as foreigners, it often induces symptoms ranging from chronic inflammation to severe device rejection. The material surfaces become befouled if minerals, proteins, platelets, bacteria, macrophages, and inflammatory tissue deposit and adhere to the surfaces.

9. Requirements for Biomaterials

All biomaterials intended for use in contact with living systems must meet certain criteria and regulatory requirements. The minimum requirements include the following: (1) the material must be biocompatible, such as nontoxic, blood- or tissue- compatible, noncarcinogneic, etc.; (2) the material must not leach or release harmful components into the living system; (3) the mechanical and physical properties of the material, such as strength, elasticity, durability, stability, etc., must be appropriate for the intended application; and (4) the desired mechanical properties must last for the expected life of the implant; (5) the materials must be sterilizable (Wallin et al., 1995; Pizzoferrato et al., 1995).

10. Sterilization Methods
All materials for biomedical applications must be sterilizable. Table 2.1 lists the commonly used sterilization methods (Yoda, 1998).

Table 2.1 Sterilization methods

Process	Agent/Condition	Remarks
Dry-heat	20 min to 6 h at 160 – 170℃	Not widely used
Steam autoclaving	Saturated steam 3 min at 110 – 160℃	Kill microorganisms by destroying metabolic and structural components essential for their replication; not suitable for materials with low softening point
Gaseous sterilization	Ethylene oxide at reduced pressure or its mixture with an inert gas at elevated pressure	Kill microorganisms due to alkylation of amine group on the nucleic acid; low temperature process suitable for a wide range of low-melting temperature polymers
Liquid sterilization	Formaldehyde solution hydrogen peroxide solution	Complete removal of formaldehyde solution is required Complete removal of cytotoxic residuals is required
Gamma irradiation	At doses of 2.5 – 5.0 Mrad	Kill microorganisms by utilizing radiation, either gamma rays from Cobalt-60 or accelerated electrons; negligible thermal effect, high efficiency, ease of packaging and sealing; inducing cross-linking and degradation

2.1.3 Applications of Biomaterials

The materials that are widely used as biomaterials include metals, ceramics, carbons, modified natural biomolecules, synthetic polymers, and composites. In a practical sense, biomaterials are manufactured or processed to be suitable for use in or as a medical devices that comes into intimate contact with biological systems. The contact is generally implemented in several ways: (1) permanent implantation (e.g., heart valves, total joint replacement, dental restoration, and intraocular lenses); (2) long-term application (e.g., fracture fixation devices, contact lenses, removable dental prostheses, hemodialysis systems); (3) transient application (e.g., needles for vaccination or phlebotomy, wound healing devices, cardiopulmonary bypass and cardiac assist systems).

The biomaterial selection criteria are determined by the specific application. Table 2.2 summarizes some important applications of implants as well as the

Table 2.2 Biomaterial Applications

Application	Biomaterials	Remarks
Cardiovascular implants		Materials must be non-thrombogenic
Heart valves	Pyrolytic carbon, Co – Cr alloys, titanium	Replace diseased heart valves
Pacemakers	Stainless steel, silicone rubber	Maintain heart rhythm
Vascular grafts	Polyurethanes, silicone rubber, Teflon, Hema-coated polymer, heparin coatings	Cure vascular disease
Stents	Ti, Ta, shape memory alloys (NiTi)	Open tiny arteries
Blood oxygenators	Polycarbonate and silicone	Replace the gas transfer function of the natural lungs during surgery
Dental implants		
Tooth replacement	Stainless steel, Co – Cr – Mo alloys, Ti and Ti alloy, bioglass	Replace diseased, damaged, or loosened teeth
Soft tissue implants		
Breast prostheses	Silicones	Augment or redefine the tissue
Artificial skin: nose, chin other prostheses	Hydrogels, Collagen	Replace lost soft tissue
Ophthalmic implants		
Contact lenses	Hydrogels, silicones, acrylics	Improve vision
Intraocular lenses	Hydrogeles, silicones, acrylics	Replace natural lens with degraded vision
Orthopedic prostheses		
Hips, Knees, Shoulder	316L stainless steel, Co – Cr alloys, Ti and Ti – Al – V alloy	Reconstruct and repair bony defects
Fixation	PMMA	
Biotechnology		
Sensors	Semiconductors	Diagnostics
Implanted batteries	Metals and semiconductors	Drive implant devices
Electronic controller		
Electrodes	Pt, Ti, Au, Ta	Diagnostics, protein immobilization, and cell function
Lab-on-a-chip	Glass, semiconductor, metals	Diagnostics

most common materials used to manufacture the implants (Christ et al., 1995; Helmus and Tweden, 1995; Peppas, 1994; Liu and Jin, 1997).

1. Cardiovescular Implants

Cardiovascular implants usually involve interactions of both blood and soft tissue with materials. These materials form the components of (1) extracorporeal devices that remove and return blood from the body, (2) devices that are inserted into a blood vessel, or (3) devices that are permanently implanted. The materials for cardiovascular applications must be non-thrombogenic.

2. Dental Materials

Dental implants require inert materials that can be prepared by relatively inexpensive and highly reproducible methods and exhibit high mechanical stability. For these reasons, highly cross-linked polymers are extensively used as denture bases, crowns, and bridges, orthodontic appliances, and artificial teeth. Additionally, such polymers must be comparable in properties with tooth enamel, adhere to the tooth enamel or tissue to which they are being attached, and not degrade or exhibit yellowing upon aging.

3. Soft Tissue Implants

For soft tissue device applications, the materials are typically implanted into soft tissue to augment or redefine the tissue (e. g., breast implants, or facial implants). Elastomers are predominantly used in the applications that require compliance with soft tissue.

4. Orthopedic Implants

In orthopedic applications, the materials are components of structural implants (e. g., hip prostheses) or are used to repair bony defects (e. g., bone screws and plates). Current orthopedic joint implants are mostly made of metallic alloys, ceramics, composites, or a combination of them. The material strength and the bond between the implants and human bone are critical factors for the proper function of implants.

5. Ophthalmic Implants

Materials used for intraocular lenses must be transparent to visible light and capable of resolving images to a minimum level when they are fabricated into appropriate lens bodies. The refractive index of the lens needs to be such that the resultant lens design will have sufficient refracting power while being able to fit within the length of the incision desired and the dimensions of the intraocular spaces. Materials must not contain any degradable components that are potentially leachable into the intraocular environment. The lens materials must also remain stable to the degradative agents that exist in the eye such as enzymes, aqueous media, and ultraviolet light. Materials also have to be strong enough to resist the damage that may be caused during surgery and handling.

2.2 The Host Responses to Materials

When a biomaterial is exposed to a living organism, there is a natural tendency for the living organism to respond to this foreign object. Many interrelated local and systemic reactions may occur at the biomaterial-biological system interface. The interactions between a host and an interface are extremely complex. The biomaterial-host interactions include, for example, coagulation, immune surveillance, healing, inflammation, mutagenicity, and carcinogenicity. These interactions are critical to the ultimate success (or failure) of implanted materials and devices. This section will discuss some key aspects of the host response to biomaterials.

2.2.1 Protein Adsorption

When an implanted material or device is in contact with biofluid in the body, a layer of proteins from the fluid adsorbs on the solid surface within seconds. This initiates the interaction between the material and the biological system.

Proteins are biological macromolecules constructed for specific and unique functions. They are high molecular weight polyamides produced by the specific co-polymerization of up to 20 different amino acids. The amino acid sequence, called the primary structure, is unique and specific to each particular protein. Protein molecules are made up of chains of amino acids linked together through so-called peptide bonds into one or several long chains. The hydrogen bonding characteristics of the polyamide bond in the backbone of the protein results in various secondary structures, such as the well-known α-helix and β-sheet. The α-helix is folded further into a more compact tertiary structure kept together by, for instance, disulfide linkages between amino acids in different parts of the α-helix. These units may further be bound together into still larger units by interchain bridges. The secondary and tertiary structures are of fundamental importance for the properties of the native proteins. During protein synthesis, the chain folds up, which gives the finished molecule its characteristic shape. The precise shape of a protein molecule enables it to do a particular job. The body makes proteins by following the decoded instructions held in genes (Creighton, 1993; Voet, 1995).

Figure 2. 1 is schematic illustration of biological system-material surface interactions. Surface crystal structure, hydrophobicity, roughness, and composition, and ions in solution all affect protein adsorption, and the adsorbed proteins would then dictate subsequent cellular reactions. In this sense, surfaces can be engineered for specific protein adsorption and to further manipulate the cellular material interactions. The study of the protein adsorption characteristics of biomaterials used for implants and devices is essential in understanding the

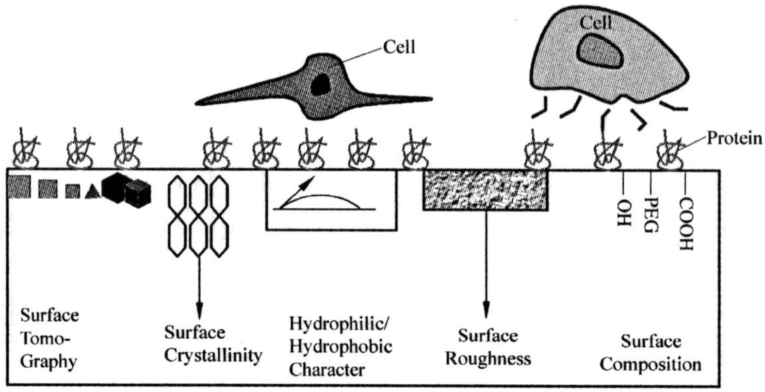

Figure 2.1 Schematic illustration of biological system – surface interactions

biological response to those materials. The subject of protein adsorption will be discussed in detail in Section 2.5.

2.2.2 Cell Adhesion

Upon immersion of an implant in body fluids, a protein film forms on the implant, and cells migrate to the surface by diffusive and active processes. The cells can adhere to material surfaces, release active compounds, recruit other cells, or grow. After cells arrive and attach to surfaces, they may multiply and organize into tissues. Among all these active processes, cell adhesion is the first event. Cells will adhere strongly to some materials, and not to others. This is determined by the special structure of individual cell membranes and materials surface properties.

A cell membrane is composed of a double layer of phospholipids embedded with various proteins, glycoproteins, lipoproteins, and carbohydrates. Different regions of the cell membrane correspond to different functions, such as absorption, secretion, fluid transport, mechanical attachment, and communication with other cells. Three main adhesive sites on cell membranes are responsible for cell – cell and cell – material surface interactions. They are adhering junctions (desmosomes), tight junctions, and gap junctions (Schakenraad, 1996; Brash, 1983). Adhering junctions hold cells together mechanically and are associated with intracellular fibers of the cytoskeleton. Adhering junctions distribute mechanical shear force throughout the tissue and to the underlying matrix by virtue of their association with intermediate filaments crossing the interior of the cell. The linkage of these filaments to the desmosomes or through these junctions to adjacent cells provides a nearly continuous fibrous network throughout an epithelial sheet.

The tight junctions also hold cells together, but they form a nearly leak-proof intercellular seal by fusion of adjacent plasma membranes, creating a continuous barrier through which molecules cannot pass. The membranes are fused by melding their lipid layers that repel large and water-soluble molecules. In invertebrates, this function is provided by separate junctions, in which the proteins of the membrane rather than the lipids form the seal.

Gap junctions allow communication between adjacent cells via the passage of small molecules directly from the cytoplasm of one cell to that of another. Molecules that can pass between cells coupled by gap junctions include inorganic salts, sugars, amino acids, nucleotides, and vitamins, while large molecules such as proteins or nucleic acids can not pass. Gap junctions are crucial to the integration of certain cellular activities. For example, muscle cells generate electrical current by the movement of inorganic salts. If the cells are coupled, they will share this electrical current, allowing the synchronous contraction of all cells in the tissue. This coupling function requires regulation of molecular traffic through the gaps.

Many mechanisms have been proposed to explain the mechanism of cell – surface adhesion. Two common ones are from the macroscopic and atomic viewpoints (Metcalf and Dalton, 1994). The macroscopic viewpoint suggests that cell adhesion depends upon physiochemical properties of the surfaces, and charges or hydrophobicity (Harris, 1973). The atomic viewpoint inserts that all adhesions are basically receptor – ligand type interactions. The receptor – ligand theory is widely accepted. A schematic of cell – surface interaction by the receptor – ligand theory is shown in Fig. 2. 2. The substrate is first bound by adhesive proteins which generally have an RGD sequence (arginine, glycine, and aspartic acid). Embedded in the plasma membrane of each cell are a number of integral proteins that interact with the surface. These proteins enable cells to "recognize" and adhere to surfaces by binding the RGD sequence to form cross-bridges which induce the change in the cytoskeleton of cells and then the morphology of cells.

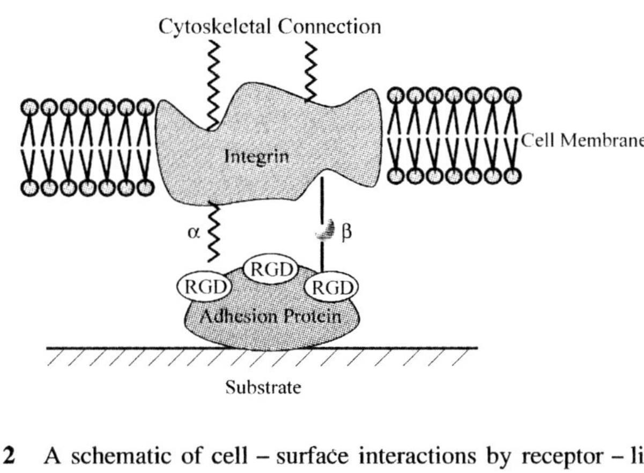

Figure 2. 2 A schematic of cell – surface interactions by receptor – ligand theory

2.2.3 Blood Clotting/Coagulation

Understanding the mechanisms of blood – material interactions is important for use of any blood contacting materials and devices. Plasma proteins and platelets are described here in detail since they are the key factors in causing the blood to coagulate and determining material – blood compatiblity. Blood mainly contains plasma proteins, red and white cells, and platelets. In blood, the packed cell volume is about 40%, and the plasma volume is about 60%. In a man, plasma is a straw-colored liquid, the main components of which are water (90% – 92%) and proteins (6% – 8%). Plasma also contains various dissolved substances, including salts, nutrients (glucose, fats, and amino acids), carbon dioxide, nitrogen wastes, and hormones. Plasma has several functions, including (1) serving as a transport system and medium for nutrients, waste products, and blood cells; (2) helping to maintain blood pressure; (3) distributing heat equally throughout the body; and (4) keeping a steady acid – base balance in the bloodstream and body (Black, 1981; Peters, 1996).

2.2.3.1 Albumin

In a plasma protein system, serum albumin occupies a position of paramount importance because of its relative abundance, osmotic and transport functions, and draws great interest for therapeutic uses. Albumin is the most abundant protein in the bodies of all vertebrates. It constitutes 50% – 60% of the total protein in blood (Black, 1981; Peters, 1996). Albumin has several important functions. Perhaps the most significant of these is its high colloid osmotic effect. Albumin is responsible for 80% of the colloid osmosis in blood, this being determined by the low molecular weight (66,000) of the protein (Peters, 1996). Thus, albumin is vital for the suspension stability of blood. Albumin is one of the most soluble plasma proteins. Solutions containing up to 30% albumin can be prepared. The high solubility of albumin is probably connected with the large number of ionizable groups present in the molecule. At a neutral pH the number of charged groups is roughly 200. Albumin has many applications, both in clinical medicine and basic research. Albumin has been widely used in surgery and in the treatment of shock and trauma. Albumin has also been used as a general protein model in a great number of studies such as amphoteric properties, denaturation processes, conformational changes, and ion bonding properties.

2.2.3.2 Fibrinogen

Of the numerous proteins in blood plasma, fibrinogen occupies a special niche with respect to blood-material interactions. Fibrinogen plays a vital role in blood coagulation and platelet aggregation. These two processes usually happen simultaneously, resulting in a platelet – fibrin plug. The blood concentration of this protein is about 3%. The total molecular weight of fibrinogen is about 330,000 (Putnam, 1975).

Another important property of fibrinogen is its ability to bind to cells such as platelets, endothelium, and leukocytes. Because of its cell adhering propensity, fibrinogen is categorized as a cell-adhesive protein. Like other adhesive proteins, fibronectin and vitronectin, fibrinogen also contains the RGD sequence. Fibrinogen is a "sticky" protein and has a strong tendency to adsorb onto various surfaces. These properties earn fibrinogen the reputation of causing thrombogenesis on biomaterials.

2.2.3.3 Immunoglobulin (IgG)

A third major class of plasma proteins is the globulins. Three types of globulin have been identified-alpha, beta, and gamma. Alpha and beta globulins are molecules that specialize in the transport of lipids, steroids, sugars, iron, copper, and other minerals, and free hemoglobin. The gamma globulins are antibodies, also called IgG, which protect the body against microorganisms and toxins. Gamma globulins are formed in the lymphoid tissues by B lymphocytes, or B cells. B cells are white blood cells that can recognize antigens through receptor molecules on their surfaces. This binding of B cells to the antigens, stimulates B cells to multiply into clones that can produce thousands of antibody molecules per second. These antibodies then attack and neutralize the invading antigens.

Antibodies attack antigens by binding to them. The binding of an antibody to a toxin, for example, can neutralize the poison simply by changing its chemical composition. By attaching themselves to some invading microbes, other antibodies can render such microorganisms immobile or prevent them from penetrating body cells. In other cases, the antibody-coated antigen is subject to a chemical chain reaction with a complement, which is a series of proteins found in blood plasma. The complement reaction can either trigger lysis of the invading microbe or attack the microbe.

Each antibody molecule is essentially identical to the receptor molecule of the original B cell that produced it. Figure 2.3 shows a schematic drawing of the molecular structure of IgG. It is composed of two pairs of polypeptide chains that form a flexible Y-shape. At the tips of the arms of the Y are the antigen-binding sites, the atomic structure of which fits the structure of the corresponding antigenic determinants. The antigen-binding site of each receptor can fit and bind to only one antigenic determinant. Therefore, the multitude of antibodies patterned after that receptor helps provide immunity, specifically against antigens with that determinant. Fc and Fab are the stem region and antigen-binding region, respectively. Two heavy chains (H) and two light chains (L) are linked to each other by interchain disulfide bonds. The broken lines in the figure represent variable portions, and the solid lines represent constant portions of the chains. Specific sites that bind antigens are formed by the variable portions.

While antigen-binding sites vary greatly among antibodies, the other regions of the molecules are relatively constant. Antibodies are grouped into five classes

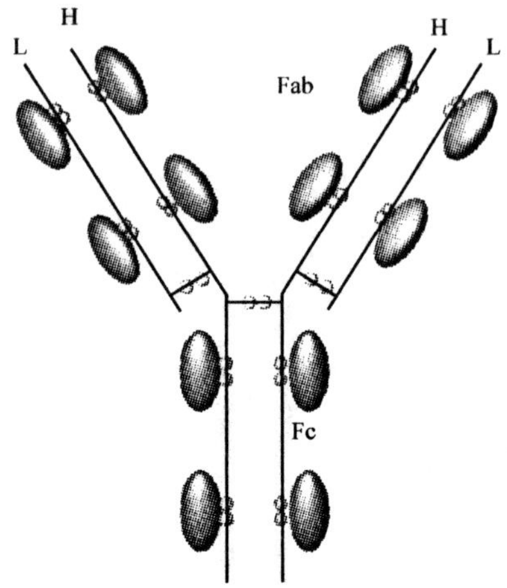

Figure 2. 3 A schematic drawing of the molecular structure of IgG, showing two pairs of polypeptide chains that form a flexible Y-shape

according to these constant regions, each class being designated by a letter attached to an abbreviation of the word immunoglobulin: IgG, IgM, IgA, IgD, and IgE. The classes of antibodies also differ in activity. For example, IgG, the most common antibody, is present mostly in the blood and tissue fluids, while IgA is found in the mucus-membrane lining of the respiratory and gastrointestinal tracts.

2.2.3.4 Platelets

Platelets are not cells, but chips of cells about 2 – 3 μm in diameter. Microscopically, they appear as flat disks and remain viable for 9 – 12 days in man. Although much more numerous (150,000 to 400,000 per cubic millimeter) than the white cells, platelets occupy a much smaller fraction of the volume of the blood because of their relatively minute size. Platelets are self-sufficient laboratories, containing all the necessary ingredients needed for adhesion, aggregation, and the formation of thrombi. When seen in fresh blood, they appear spheroidal, but they have a tendency to extrude hairlike filaments from their membranes. They adhere to each other but not to red cells or white cells. Platelets require storage at 20 – 24℃ with continuous agitation to maintain adequate function (Wood et al., 2000).

The function of the platelets is related to hemostasis, the prevention and control of bleeding. When the endothelial surface (lining) of a blood vessel is injured, platelets in large numbers immediately attach to the injured surface and to each other, forming a tenaciously adherent mass of platelets. The effect of the platelet response is to stop the bleeding and to form the site of the developing blood clot, or thrombus. If platelets are absent, this important defense reaction cannot occur, and protracted bleeding from small wounds (prolonged bleeding time) results. The normal resistance of capillary membranes to leakage of red cells is dependent upon platelets. A severe deficiency of platelets reduces the resistance of capillary walls, and abnormal bleeding from capillaries occurs, either spontaneously or as the result of minor injury. Platelets also contribute substances essential for normal coagulation of blood, and they cause shrinking, or retraction, of a clot after it has been formed. Additional information on thrombosis can be found in blood clotting.

Platelets are formed in bone marrow by segmentation of the cytoplasm (the cell substance other than the nucleus) of cells known as megakaryocytes, the largest cells of the marrow. Within the marrow, the abundant granular cytoplasm of the megakaryocytes divides into many small segments that break off and are released as platelets into the circulating blood. There are no reserve stores of platelets except in the spleen, in which platelets are present in higher concentration than in the peripheral blood. Some platelets are consumed in exerting their hemostatic effects, and others, reaching the end of their life span, are removed by reticuloendothelial cells (any of the tissue phagocytes). The rate of platelet production is controlled, but not so precisely as the control of red cell production. A hormonelike substance, thrombopoietin, which has not been identified chemically, is believed to be the chemical mediator that regulates the number of platelets in the blood by stimulating an increase in the number and growth of megakaryocytes, thus controlling the rate of platelet production.

2.2.3.5 Blood Coagulation Cascade

When blood comes into contact with foreign substances, two classes of components are affected, namely, blood cells and plasma proteins, including coagulation factors. Blood coagulation has been one of the major problems in manufacturing biomedical microdevices on a micro- and nano-scale. Blood coagulation proceeds through a "cascade" of reactions. At least 12 plasma proteins interact in a series of reactions leading to blood clotting. Figure 2.4 shows how fibrin is formed in response to three independent mechanisms: the intrinsic pathway shown on the left, the platelet activities shown in the middle, and the extrinsic pathway shown on the right (Szycher, 1983). The extrinsic pathway is activated by the blood exposed to a variety of substances naturally found in tissues, known collectively as "tissue factors". Tissue factors include collagen, basement membrane constituents, lipids, and proteins.

The sequence is initiated by the activation of a single factor, either by Factor VII of the extrinsic pathway, when platelet adhesion occurs with release of biochemical contents of their granules, such as phospholipids, or by the activation of Factor XII of the intrinsic pathway. The activated factor then reacts with other factors, sequentially, until the final step when the thrombus acts on

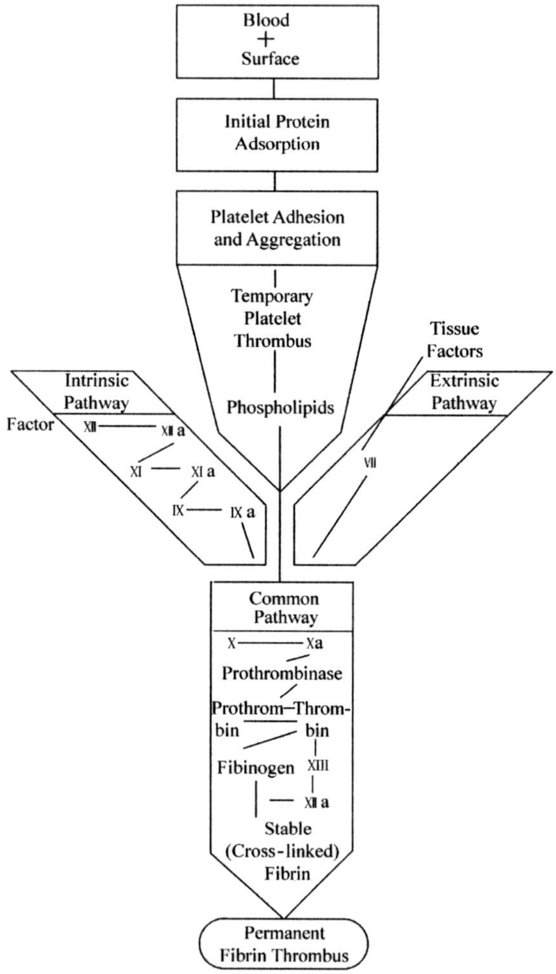

Figure 2. 4 Blood coagulation with both the intrinsic and extrinsic systems and their common path

fibrinogen to form a fibrin meshwork. The sequence appears to follow a cascade, and for that reason, this chart is also known as the blood coagulation cascade.

The intrinsic pathway plays a major role in blood – material interactions. The exposure of blood to external surfaces leads in many cases to the adsorption of Factor XII upon the surfaces, with the subsequent deposition of platelets and leukocytes. Initiation of clotting occurs either intrinsically by surface-mediated reactions or extrinsically through factors derived from tissues. The two systems converge upon a final common path, which leads to the formation of an insoluble fibrin gel when the thrombin acts on fibrinogen. The thrombus is frequently associated with fibrin formation, and the presence of fibrin stabilizes the thrombus.

2.2.4 Tissue Response

2.2.4.1 Inflammation and the Foreign Body Reaction

Inflammation is generally defined as the reaction of vascularized living tissue to a local injury (Anderson, 1998). The foreign body reaction involves parts of this response with particular regard to an implanted device. Immediately after injury, there are changes in vascular flow, caliber, and permeability. Fluid, proteins, and blood cells escape from the vascular system into the injured tissue. After changes in the vascular system, cellular events occur and characterize the inflammatory response. Regardless of the tissue or organ into which a biomaterial is implanted, the initial inflammatory response is activated by injury to vascularized connective tissue. Since blood and its components are involved in the initial inflammatory responses, blood clot formation and/or thrombosis also occurs.

The inflammatory response is a series of complex reactions. The predominant cell type present in the inflammatory response varies with the age of the injury. Neutrophiles, which are the characteristic cell type of acute inflammation, predominate during the first several days after the implantation. After the acute inflammatory response, monocytes and lymphocytes predominate in the implant site and are the characteristic cells of chronic inflammation. The monocytes in the chronic inflammatory responses, migrating from the blood, differentiate into macrophages within the tissue at the implant site. These macrophages will fuse or coalesce into giant foreign body cells.

2.2.4.2 Wound Healing and the Repair of Implant Sites

Wound healing and the repair of implant sites involve two distinct processes: regeneration, which is the replacement of injured tissue by parenchymal cells of the same type, and replacement by fibrous connective tissue that forms a capsule (Ratner, 1996; Silver, 1994). These processes are generally controlled by either the proliferative capacity of the cells in the tissue or organ receiving the implant. With regard to the regenerative capacity of cells, there are three groups of cells: labile, stable, and permanent cells that cannot reproduce themselves after the birth of the host.

Perfect repair, with a restitution of the normal structure, theoretically occurs only in tissues consisting of stable and labile cells. Tissues composed of permanent cells (e.g., nerve cells, muscle cells, cardiac muscle cells) most commonly undergo an organization of the inflammatory exudate leading to fibrosis capsule formation (Silver and Christiansen, 1999). Tissues composed of stable (e.g., parenchymal cells of the liver, kidney, and pancreas), mesenchymal cells (e. g., fibroblasts), and vascular endothelial and labile cells (e. g., epithelial cells and lymphoid and hematopoietic cells) either form fibrosis or undergo resolution of the inflammatory exudate, leading to the restitution of the normal tissue structure.

Following injury, cells may undergo adaptation of growth and differentiation. Important cellular adaptations are atrophy, hypertrophy (an increase in cell size), hyperplasia (an increase in the cell number), and metaplasia (changes in cell type). Hyperplasia of smooth muscle cells at blood vessel/vascular graft anastomoses may lead to a failure of the graft by stenosis or occlusion, i. e., narrowing of the lumen, and thrombosis. Other adaptations include a change in which cells stop producing one family of proteins and start producing another, or begin a marked overproduction of protein. This may be the case in cells producing various types of collagens and extracellular matrix proteins in chronic inflammation and fibrosis. Causes of atrophy may include a diminished blood supply and inadequate nutrition caused by fibrous capsules surrounding the implants.

2.2.4.3 Fibrosis and Fibrous Encapsulation

Implanted medical devices in vascularized tissue are commonly encapsulated in a fibrous capsule (Anderson, 1998; Neville, 1993). Fibroblasts play a critical role in fibrosis. Collagen-specific receptors are activated when fibroblasts adhere to implant surfaces. This results in collagen deposition, which subsequently forms a fibrous layer surrounding the capsules. This suggests that the provisional matrix at medical device/tissue interfaces does not undergo reorganization to recapitulate the tissue into which the device has been implanted, but instead produces the chronic response of fibrous capsule development.

The formation of a fibrotic tissue layer around cellular encapsulation devices has been a limiting factor in their therapeutic effectiveness and in the overall success of *in vivo* operations. Fibroblasts play an important role in the deposition of a fibrous layer surrounding an implant. Following adhesion of the fibroblast to an implant, collagen-specific receptors are activated which may result in collagen depositions and subsequent walling in of the capsule implant by the deposition of a fibrous layer. Excessive fibrosis would prevent the cells within capsules from obtaining their nutrients, leading to the death of the cells and the failure of the biocapsule device. Although a fibrotic response occurs to some extent with all implants, the thickness of the fibrotic layer largely depends on the surface properties of the implant. Therefore, it is of great interest to improve material tissue biocompatibility through the reduction of fibroblast attachments by changing the material surface properties.

2.2.4.4 Infection

Bacterial adhesion and colonization onto biomaterial surfaces and the subsequent infections are frequent causes for the failure of many medical devices and implants (Park et al., 1998; Gristina et al., 1985).

In general, there are four main types of infections associated with implants. First, infections may occur when sterilization techniques sometimes are ineffective in creating conditions of total sterility. Second, the superficial immediate infection is due to the growth of organisms on the skin (or near it). Examples of this include suture infections and growth of microorganisms under burn dressings. These infections can usually be traced to bacteria residing normally on the skin or to airborne bacteria that are trapped and cultured in the moist conditions at the superficial site. Third, from deep infections occurring immediately after surgery. The bacteria are most usually skin dwelling types carried into the implant site during the surgical procedure. The fourth source comes from the physical disruption of tissue during surgery, resulting in the release of material in cysts or sequestered sites into areas where the bacteria experience better growth conditions. Once the bacteria adhere, they can multiply, form complex multilayered colonies, and produce a slimy matrix material that encases the bacterial cells, called a biofilm. The formation of a biomaterial-associated biofilm usually leads to the removal or revision of the affected device or implant. The mechanism of bacterial adhesion is a very complicated topic (An and Friedman, 1998). The key to biofilm formation appears to be the interactions between the body and the implant, more specifically, the interface between the biomaterial surface and the bacteria as well as the associated environments. For example, plasma proteins deposited onto the implant material surface can "condition" the surface for biofilm formation. The surface characteristics and properties of a biomaterial, such as the roughness and area, hydrophobicity, porosity, and chemistry, have a significant effect on bacterial adherence and colonization. It is important to advance our understanding of the mechanisms of bacterial adhesion and prosthetic infection in order to attain appropriate methods for preventing infections.

2.3 Biological Performance of Biomaterials

2.3.1 Swelling and Leaching

When implant materials contact a biological system in the absence of reactions, the materials transfer across the material – tissue interface. If the substance, primarily fluid, moves from the tissue into the biomaterial, the volume of the material increases due to the conservation of volume. This phenomena is called swelling. If the components of the biomaterial dissolve in the tissue, the materials become porous. This is called leaching. Swelling and leaching result from the process of diffusion. Both of these effects have profound influence on the behavior of materials. For example, swelling and leaching might cause a large deformation in materials.

Swelling and leaching both affect a material's mechanical properties. Swelling produces continual deformation, which may eventually lead to a mode of failure. Excessive leaching, such as the intergranular leaching in metals, can produce defects that coalesce into macroscopic voids and result in a reduction in fracture strength.

2.3.2 Corrosion

Electrochemical phenomena play an important role in the performance of metallic biomaterials. When a metal is introduced into the body, a wide variety of processes and interactions with the biological environment can take place. Metallic biomaterials, as a general group, are susceptible to corrosion in human bodies by biomechanical and biological factors. There are many forms of corrosion, and the most common ones are galvanic, crevice, fretting (fatigue corrosion), pitting, intergranular, and stress corrosion (Black, 1981; Williams et al., 1995).

1. Galvanic Corrosion

Galvanic corrosion takes place when two metals are physically in contact and are immersed in a conducting fluid medium. If two metals (plate and screw) are independently placed within the same biological fluid, each has its own electrochemical potential with respect to the solution. The one with a positive charge becomes anodic and the other becomes cathodic. If the two metals make physical contact, electrons will flow passing from the anodic metal to the cathodic metal. This causes a continuous process of dissolution of the anodic metal. The corrosion usually can be judged by the discoloration observed during the routine device removal from implanted tissue sites.

2. Crevice Corrosion

Crevice corrosion is initiated by crevices presented in material structures. The crevice can be either an interface between parts of a device, or a defect such as an incomplete mechanical crack. The details of the initiation of crevice corrosion are not yet clear. Once begun, however, it is characterized by oxygen depletion in the crevice, anodic metallic corrosion along the crevice faces, and cathodic protective conditions on the metal surface around its crevice. Static nonflowing conditions in the solution seem to favor crevice corrosion.

3. Fatigue Corrosion

Many metals display limited endurance in regard to cyclic loading. The maximum stress that can be reached without failure continuously decreases as the number of load cycles increases. The phenomenon is termed fatigue corrosion, which is a critical factor in determining the life of metallic implants undergoing cyclic mechanical loading. For instances, materials for use in replacing diseased heart valves must be selected for fatigue corrosion resistance.

4. Pitting Corrosion

This is a special case of crevice corrosion. It is caused when localized corrosion occurs as a result of imperfections of the passive protection layers on metals. Many metal implants are passivated by a very thin layer of oxide. In the cases where the passive layer is broken down, the metal usually does not repassivate, and active corrosion can occur. These localized spots will actively corrode, and pits will form on the surface of the material. If this is combined with galvanic corrosion, it will result in a large degree of localized damage since the small areas of active corrosion become the anode and the entire remaining surface becomes the cathode.

5. Stress Corrosion

This occurs when an applied stress and a corrosive environment work together to cause complete failure of a component, while neither the stress nor the environment alone would cause the failure. Since the formation of small cracks in a static and loaded structure will cause stress concentration, this will lead to rapid growth of cracks and eventually structural failure. Pits are a hazard in highly stressed implants as they constitute points of stress concentration and may serve as the starting points for mechanical cracks to develop.

2.3.3 Particulates

Implants will release materials by dissolution or release of particulates when they are exposed to biologic fluids (Nagase, 1995; Raay et al., 1995; Syrjanen et al., 1986). Particulate-wear debris is now known to play a central role in periprosthetic bone resorption and prosthetic loosening (Schmalzried et al., 1993). Host responses to particulates are the same as that to bulk materials except that particulates have a much larger surface area to interact with the surrounding tissues. Therefore host reaction to particulates can take several forms: inflammation, immune reaction, mitogenic reaction, tumor formation, and toxic reaction. The surface structure, crystallinity, surface charge, and the size of particulates will all affect the cell – particle interactions.

Many experiments found that the size of the particle is a critical factor which induces harmful effects on hosts as compared to bulk materials (Pratten and Lloyd, 1986; Shanbhag et al., 1994). It has been suggested that the harmful effects depend more on the particulate nature of the material than its chemical biocompatibility. It was found that all particles of less than 5 – 10 μm diameter would cause cell damage by direct contact despite the chemical nature of the material. However, there is not yet a general understanding of the mechanisms by which particulate materials might exert a harmful effect greater than that of their parent material. Thus, the design of orthopedic implants to avoid production of fine particulate debris has become more important than the search for materials of better biocompatibility.

2.3.4 Surface Topography

Many studies have shown that surface topography, including surface geometry, orientation, porosity, roughness, curvature, edges, and confined spaces, is an important factor in determining cell attachment, adhesion, orientation, and function (Curtis and Wilkinson, 1997). The surface topography of a substrate markedly influences the cell shape that can regulate cell growth, cytoskeleton gene expression, and differentiation (Chou et al., 1995). However, little is known about the mechanism by which surface topography exerts its effects, although many studies have focused on this subject. Surface topography has not been fully exploited as a tool to engender desired biological responses to implanted devices, and there is considerable potential in applying surface topography in innovative ways to improve implant performance.

2.3.4.1 Surface Roughness

Many researchers have reported that connective tissue cells, endothelium, epithelium, peripheral nerves, and bone cells attach to rough surfaces move easily than to smooth surfaces of various polymers and metallic implants (Curtis and Wilkinson, 1997; Bobyn et al., 1982; Pilliar, 1983). One explanation of this phenomenon is that a rough surface would increase the effective surface area for cell attachment and surface roughness promotes mechanical interlocking by tissue ingrowth. It has also been reported that glass, metal plastics, and silicon-rubber implants with smooth surfaces induce sarcomas in rodents whereas the same respective materials with rough surfaces do not induce sarcomas (Boone et al., 1979). One hypothesis emphasizes the role of the connective tissue capsule in causing foreign-body cancer formation. It suggests that fibroblasts and collagen bundles respond to smooth subcutaneous implants by aligning parallel with the implant surface to form a capsule. A second hypothesis stresses the role of macrophages and giant cells that accumulate on rough, but not on smooth surfaces, in inhibiting sarcoma induction. This view suggests that cancer formation may be due to the paucity of macrophages on the smooth surfaces. However, when implants are used as in percutaneous or dental implants, or come in contact with bone marrow or cells of the immune system, smoother surfaces are favored, for they attract fewer microorganisms and produce fewer immune-related responses. The rough surfaces can be created by rubbing the surface with abrasive particles, blasting the surface with metal particles, or etching the surface randomly with ion beams.

2.3.4.2 Porous Surfaces

Porous surfaces have been used in percutaneous, subcutaneous, orthopedic, and dental implants to promote tissue interlocking. One explanation is that epithelial cells interlock into the porous surfaces and organize adjacent tissues. Variables such as pore size, percent porosity, fiber diameter, and surface roughness of the matrix determine the speed and completion of ingrowth (von Recum and Park, 1981). Many investigators examined the effect of pore size and porosity on soft tissue healing in a number of different natural and synthetic polymeric systems and found that pore size must be large enough to allow blood vessel ingrowth (at least 40 μm). However, if the pore sizes are too large, the ingrowth effect decreases. It was found that the optimal pore size for collagen based wound dressings is approximately 100 μm.

However, interconnected poroues epithelium-penetrating devices would also cause problems. In a review on percutaneous devices, von Recum and Park noted that implant materials with porous surfaces, including open sponges, felts, and velours, are especially prone to serious implant infection that will never reheal, possibly because the pores provide an easy pathway for microbial invasion. Feldman et al. (1983) assert that percutaneous devices with interconnected pores never completely fill with tissue, and, instead, they fill with intercellular fluids. Thus, such surfaces may serve as wicks and allow bacteria free access to subepidermal tissue through the interconnecting porosity. Similar observations have also been reported with dental implants and have led to the notion that porous surfaces should not be used in the region of epithelial attachment to dental implants.

Porous surfaces can be thought of as related to rough surfaces. The difference between the two is that the porous surfaces have deeper, channel-like surface irregularities (10 – 1000 μm), and the channels or pores are interconnect as compared to rough surfaces. Typically, porous surfaces are created by sintering metal beads, plasma spraying metals or ceramics, weaving textiles, hydrothermal chemical exchange, and charged-ion bombardments.

2.3.4.3 Micromachined or Microetched Surfaces

Many experimental studies have demonstrated that proper material surface topography can provide guidance for the growth of various types of cells, including epithelial cells, fibroblasts, nerve axons, neutrophils, and osteoblasts. Figure 2.5 shows human gingival fibroblasts oriented and elongated in the direction of the grooves. Topographic guidance refers to the tendency of cells to be guided in their direction of locomotion by the shape of the substratum. Topographic guidance is a very robust phenomenon that can be exhibited *in vitro* by diverse cell types on diverse surfaces with diverse geometries (Chen et al., 1998; Mrksich et al., 1997).

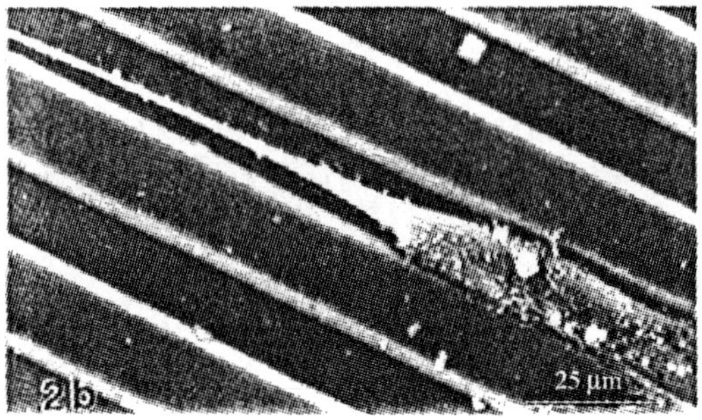

Figure 2.5 A SEM picture of human gingival fibroblasts oriented and elongated in the direction of the grooves. Photo courtesy of Chehroudi and Brunette (1995)

Detailed studies on the relationship between surface topography and cell behavior were made possible by the use of microfabrication techniques to construct different surfaces of highly ordered and precise topographies (Meyele et al., 1991; Chen et al., 1998). Micromachined surfaces can induce cell contact guidance *in vivo* that results in inhibition of epithelial downgrowth on percutaneous devices. Their different dimensions can also promote selective ingrowth of either epithelium or fibroblasts. For example, epithelial tissue tends to bridge deep grooves (Chou et al., 1995). This bridging, by which the cells contact only the ridges of the grooved surface, probably is the result of the tension formed within the epithelial sheet. In contrast, fibroblasts interdigitate into deep grooves (19 – 30 μm) to form a perpendicular arrangement with the surface similar to the functional orientation observed in the periodontal ligament of teeth. The oriented connective tissue inhibited epithelial downgrowth indirectly on the surface of percutaneous devices (Chehroudi et al., 1992). Osteogenic cells appeared to be influenced by the micro-environment provided by deep grooved or pitted micromachined surfaces (30 – 120 μm), resulting in the differentiation and expression of a mineralized matrix at the implant interface.

2.4 Biocompatibility Assessment

The interactions between a host and an artificial surface are extremely complex. The biomaterials – host interactions, as discussed before, include coagulation, immune surveillance, healing, inflammation, mutagenicity, and carcinogenicity. An ideal material for every possible application may never exist. Clearly, as the knowledge and understanding of the molecular mechanisms controlling these interactions increase, we may optimize biomaterials for intended applications. Significant opportunities and challenges exist in the characterization of the biocompatibility of biomaterials. Current methods for evaluating materials biocompatility are outlined in this section. Approaches for controlling the interface between tissue and biomaterials and ways in which the engineered materials may contribute to medicine are considered.

The biocompatibility of an implant material should be evaluated quantitatively. Biocompatility tests have undergone a change in emphasis during past decades, and now it is generally accepted that they can be grouped in two levels (Pizzoferrato et al., 1995b), as shown in Fig. 2.6. At the first level, tests are concerned with "biosafety" issues. These tests are designed to study the severe deleterious effects of biomaterials on organisms regardless of the final use of the materials. At the second level, the tests are concerned with biofunctionality issues. These tests are designed to assess whether materials only evoke an appropriate host response in a specific application. The first level tests mainly include cytotoxicity, hemotoxicity, the Ames test, and histotoxicity studies. The second level tests include cytocompatibility, immunocompatibility, hemocompatibility, histocompatibility, and infectability. It is essential to go step by step along Level 1 testing. Only materials that pass Level 1 testing should be further evaluated by Level 2 testing.

2.4.1 First Level Tests: Biosafety Testing

1. Cytotoxicity

The cytotoxicicity study is designed to rule out the acute toxicity or cytotoxicity of materials, regardless of their final use. Then, the materials that are nontoxic from first level testing are further inspected by cytocompatibility tests, a second level test.

Cells are extremely sensitive to toxic agents and their biological responses can be easily detected. Mammalian cell cultures are thereby commonly used for assessing the *in vitro* biocompatibility of materials (Johnson et al., 1985). The established cell lines, either commercially available or following the subculture of primary cells, are recommended by most cytotoxicity evaluation protocols (Wallin and Upman, 1995). The cell features to be evaluated include cell morphology, cell viability and/or cell death, cell adhesion to the material surface, and cell growth in the presence of the material. The most commonly used cell lines for cytotoxicity testing are Hela cells derived from a human cervical carcinoma or L-929 cells from a mouse fibroblast cell line.

Cell morphology and death can be examined by light (Van-Kooten, 1998), or electron microscopy (Peluso et al., 1997), and recently by means of quantitative imaging (Reid, 1993). Cell viability can be found by the uptake of a vital stain, such as neutral red, trypan blue, erythrosin B, etc., by checking the membrane permeability. The ability of a cell to adhere to material surfaces is generally considered a sign of compatibility. Cell growth on biomaterials can be

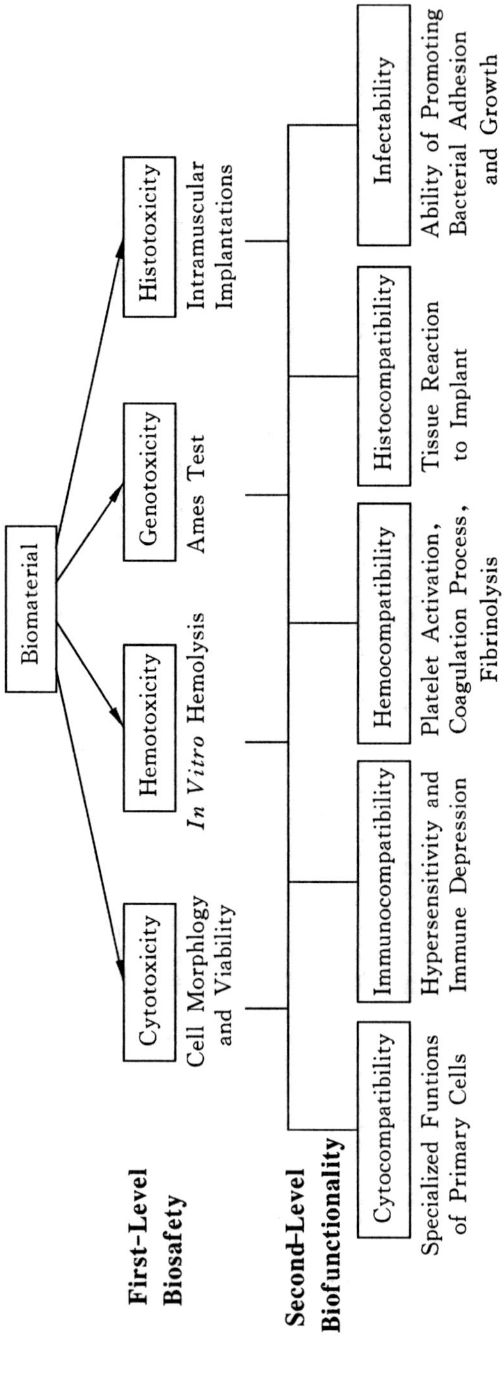

Figure 2.6 Commonly used biocompatibility tests

measured by direct cell enumeration, protein content measurement, crystal violet, or through the methylene blue assumption (Chirila et al., 1992; Takaoka et al., 1996). The validity of the assay should always be complemented by direct visual methods.

2. Hemotoxicity

Materials are screened for hemotoxicity by evaluating the damage to red blood cells. After *in vitro* contact of whole blood with materials or with its extract, the hemoglobin released is monitored using a spectrophotometer (Singh et al., 1990). This assessment is particularly relevant for the materials used in artificial heart valves and extracorporeal circuits, as these devices have been shown to some extent to cause hemolysis. This test is also appropriate for some materials that polymerize *in vivo*, when the release of volatile low molecular weight compounds into tissues is suspected.

3. Genotoxicity

A major complication with the development and implementation of biomaterial implants for medical therapies is the assessment of material-induced genotoxicity. The initial purpose of genotoxicity testing was to determine whether a material had the potential to induce heritable genetic alterations in humans. Genotoxicity can be evaluated using both *in vitro* and *in vivo* methods of different complexity and reliability (Kilbey et al., 1977). The goal for the genotoxicity test *in vivo* is to obtain quantitative data from relevant animal model systems, from which extrapolations to humans can be made for predicting the safe use of the implant material. Screening tests for the assessment of material-induced mutagenesis *in vitro* are highly desirable in the hope of replacing cost-intensive and time-consuming animal experimentation (Kajiwara et al., 1997). In the screening phase of a material, the Ames test is recommended. Materials proved positive are excluded from any further evaluation. Numerous chemicals have been studied, and it has been documented that most substances acting as carcinogens *in vivo* yield a positive result *in vitro*; conversely, most of the noncarcinogenic substances yield a negative result in the Ames test for mutagenicity.

The Ames' test is able to detect retromutations induced by materials on genetically altered (mutant) *Salmonella* strains that are unable to synthesize histidine by themselves. Exposure to mutagens can favor retromutations to the normal nutrient requirements. Consequently, retromutant *Salmonella* strains are able to grow on a histidine-free culture medium, whereas nonretromutant *Salmonella* are unable to grow. In order to determine the effect of substances with different mutagenic activity, it is necessary to use various strains (Wening et al., 1995). In addition to the mutation for histidine, each strain shows other mutations that make it sensitive to various genotoxic substances.

Solid substances can be assayed by the Ames test after being extracted into a fluid. Of particular interest is the assessment of the mutagenic capability of biomaterials such as bone cement or dental endodontic sealers that undergo an *in vivo* polymerization process. Actually polymers, more than any other material, represent a potential source of toxic substances, including monomers or residual oligomers, plasticizers, and stabilizers. In these cases, both single components and polymerized materials are tested at different end points after curing.

4. Histotoxicity

Due to the inherited limitations of cell culture systems, animal implantation is needed in order to assess the histotoxicity of materials. The implantation permits assessment of the local pathologica effects of a sample material that is surgically implanted or placed into an implant site or tissue for an intended application.

The configuration of the material to be assayed is extremely important for correct interpretation of the test results. An unknown material must always be implanted along with a reference material, with the same shape, size, and surface structure in a similar site. The materials to be tested are usually implanted subcutaneously in small animals (mainly rats) by placing the materials into a pocket created between the derma and muscle at the scapular level. The material is inserted deeply into the pocket so as not to interfere in the wound-healing process. At retrieval, the specimen is removed together with a large zone of the implant tissue.

Evaluations of histological specimens are usually qualitative. Tissue necrosis at the implant site, if not occurring at the reference material implantation site as well, is regarded as a sign of histotoxicity. Necrosis, according to the ISO document, is revealed by the presence of nuclear debris or damage to the small vessels. The materials inducing tissue necrosis are excluded from further experiments.

2.4.2 Second Level Tests: Biofunctional Testing

1. Cytobiocompatibility

Cytocompatibility tests mainly use primary cell cultures that are representative of the cell types with which implants directly contact. The most commonly used primary cells include mouse macrophages, thymocytes, fibroblasts, human peripheral blood lymphocytes and granulocytes, endothelial cells, corneal cells, and osteoblasts.

As noted previously, level 2 biocompatibility tests focus on the specific function of a medical device. Thus, human cells relevant to the function of the device should be used for individual tests. For example, endothelial cells would be the logical choice for the biofunctionality testing of vascular prostheses, platelets, monocytes and granulocytes for all blood-contacting devices, and fibroblasts, chondrocytes, and osteoblasts for studying cell interactions with orthopedic or dental implants.

2. Immunocompatibility

Most implants are capable of acting as antigens, stimulating antibody production by plasma cells and activating or deactivating T cells. Immediate hypersensitivity occurs in some animals after exposure to an antigen. A biomaterial should be immunocompatible. This means that the immune system should tolerate the implantation of a foreign material and, conversely, the foreign material should not interfere with the host's immune defenses. Immunoresponse to materials is a very complex process. Details can be found in many texts and reference books. Here we describe some basic tests that are most commonly used to assess material immunocompatibility. At present, no protocol to study immunocompatibility has been standardized. Commonly used testing methods are listed below.

- *In vitro* testing

These tests are designed to detect how lymphocytes and macrophages respond to implants (Jenny and Anderson, 1999) and usually involve preparing a soluble or colloidal extract of the material to be tested. Lymphocyte transformation testing is measured by evaluating the T lymphocyte surface antigen, lymphocyte activation by an antigen, and the cytotoxic activity of K and NK cells against target cells. Macrophage and polymorph activity are quantitatively evaluated by measuring the phagocytosis of particulated biomaterials, the chemotaxis of cells that have engulfed material particles and oxidative bursts (Syrjanen et al., 1986; Hunt et al., 1997).

- *In vivo* testing

The hypersensitivity test is most commonly used to characterize material immunocompatibility. Immediate hypersensitity is characterized by redness, swelling, and pain at the insertion site of the biomaterials within 24 h. Delayed hypersensitivity can occur 48 h after insertion. Antibody production is maximized 14 days after insertion. Fourteen days after implantation, blood is drawn from the animal, and antibody-antigen complex levels can be determined by immunoassay. In some cases, the initial immunocompatibility tests can be simplified by measuring the amount of antibody - antigen complex with an immunoassay by direct exposure of the blood to implants.

3. Hemocompatilibility

Hemocompatibility is evaluated using endothelial cell culture and changes in blood constituents induced by the materials. Hemocompatibility assessment is recommended especially for blood-contacting materials. Such materials must have no detrimental effects on platelets and coagulation factors, erythrocytes, leukocytes, proteins, enzymes, and other blood constituents.

- *In vitro* testing

Several hematologic tests have been used to evaluate blood compatibility. They place whole blood on a test surface for short legnths of time (i. e. 1 s to 1 min) or until clotting occurs at room temperature. Clotting time, clot weight, the number of adherent platelets, thrombin time, prothrombin time, partial thromboplastin time, platelet factor release, and platelet counts are measured. Short or excessively prolonged clotting times suggest that the material is either activating or deactivating the clotting system and is evidence of blood incompatibility.

The methods for assaying changes in platelet functions after exposure to biomaterials are regarded as the most sensitive. Platelet-rich plasma or suspended platelets are usually used rather than whole blood. The degree of platelet adhesion and the morphology of the attached platelets are determined. The adhesion is measured by counting platelets before and after they contact the materials. The morphologica modifications of platelets induced by biomaterials are evaluated directly by both light microscopy and quantitative imaging.

Analysis is also made on the deposition of plasma proteins (see Sect. 2.5) onto the surface of the material. A surfaces with less plasma adsorption would lead to less platelet adhesion (Hsiue et al., 1996). Protein adsorption can be measured by fluorescence spectroscopy, infrared spectroscopy, immunoassay, radioimmunoassay, radiolabeling, circular dichroism, ellipsometry, contact angle determination, and Fourier transform infrared spectroscopy. The most commonly used technique involves the measurement of ^{125}I radiolabeled proteins (albumin, fibrinogen, or gamma globulins) absorbed onto the surface of a biomaterial.

· *Ex vivo* testing

This involve the blood – material contact occurring outside the animal body. A shunt connected to the vessels is the simplest model (Silver et al., 1995). Also, these methods do not require anticoagulants. Unlike *in vivo* methods, they permit the simultaneous evaluation of different materials.

· *In vivo* testing

These tests involve the implantation of test samples into a number of different test sites and in several different configurations. The surgical procedures used to prepare animals for these types of procedures require the use of anticoagulated blood. The material to be tested can be implanted for a predetermined period of time as a vascular graft, or it can be tested in a vascular catheter or arterio-venous shunt or in a cardiac assist device. The material is then removed, and the amount of clot and platelet aggregates attached to the surface of the material is determined. In some cases, the animal's platelets are radioactively labeled and the platelet half-life is used to assess platelet survival time.

For long-term blood-compatibility evaluation, vena cava ring tests are widely used. Rings composed of or lined with the material to be tested are inserted into the vena cava of an animal or dog and removed after a predetermined period of time. The "patency" of the ring is then assessed by weighing the clot and by visually determining the fraction of the vessel that remains open. The degree of embolization can be measured by determining the amount of clot that has been filtered from the blood by the animal's kidney or lungs, when the ring is inserted into the aorta or vena cava, respectively.

4. Histocompatibility

The histotoxicity evaluation by intramuscular or subcutaneous implantations, as well as the other first-level tests, is not sufficient to validate the biocompatibility of a material. Assessment of material histocompatibility is thereby necessary. Histocompatibility assessment relies on the examination of tissues around the materials, implanted in the same shape and place as they will have *in vivo*. Depending on the final use of the device, soft tissue, vessels, bone, or nervous tissue is chosen as an implantation site.

The histocompatibility of materials is assessed quantitatively by a grading system for cellular response at animal implant sites. This includes the determination of the inflammatory cells (type and number) and the evaluation of fibrous capsule (thickness and organization) (Black, 1981).

5. Infectability

Infection is normally evaluated histologically based on observation of tissue-implant interfaces (Bellantone et al., 2000). The presence of leukocytes and positive staining for gram-negative bacteria are evidence of an infection. In some instances the samples are contaminated in a controlled manner with organisms prior to sterilization to evaluate the efficiency of sterilization. In addition, *in vitro* tests are used to evaluate whether contact with a biomaterial inhibits the ability of leukocytes to phagocytize and kill specific bacteria. Bacteria adherent to biomaterials are quantified using microbiological techniques (e.g., turbidimetry). The quantitative evaluation of bacterial adhesion has to be assessed on both controls and samples. Nylon and polystyrene are usually chosen as a negative control (does not adhere to bacteria) and a positive control (prone to bacterial adhesion), respectively.

2.5 Protein Adsorption at Interfaces

Protein adsorption onto a solid surface influences a variety of processes that occur at interfaces (Nygren and Stenberg, 1988; Walivaara et al., 1992). The biocompatibility of clinical implants, mammalian and bacterial cell growth in culture, initiation of blood coagulation, complement activation by surfaces, solid phase immunoassays, biosensor development, and protein binding to cell surface receptors all involve protein adsorption at interfaces. Practical problems such as contact lens fouling, foaming of protein solutions, and fouling of equipment in the food processing industry are direct consequences of the surface activity of proteins. In general, any process involving a surface in contact with a protein solution is likely to be influenced by protein adsorption to the interface. Despite a fairly long history of the study of proteins at interfaces, many of the fundamental mechanisms remain only partly understood, and research on proteins at interfaces remains very active. In this section, the relevant fundamental principles of protein adsorption are discussed, including driving forces for adsorption, transport mechanisms, the modes of proteins binding to surfaces, and the adsorption theory.

2.5.1 Concepts of Adsorption

Adsorption is the attraction of molecules, atoms, or ions of a substance to a solid surface. It is traditionally classified as either physical or chemical. In physical adsorption the adsorbed molecules retain their individuality, and the forces involved are similar to Van der Waals forces in nonideal gases. In chemical adsorption (or chemisorption), the adsorbed molecule forms a chemical connection with the solid surface through strong chemical interaction forces, which to some extent include ionic interactions. The adsorbed molecules, although it forms a chemical bond, can retain its individuality to some extent. Although it is similar to that of a small molecules, the adsorption of proteins is distinguished by the large size of the adsorbate and by the fact that the adsorbed protein can undergo various transformations (Adamson and Gast, 1997).

2.5.2 Protein Adsorption at Interfaces

Proteins are intrinsically surface-active since they are large and amphipathic molecules. The fundamental classes of forces and interactions important in proteins are classified into four major types: (1) ionic or electrostatic interactions due to the attraction or repulsion of two or more groups carrying net electrical charges; (2) hydrophobic interactions, a largely entropically driven process due primarily to water structure effects adjacent to hydrophobic interfaces. In this interaction, water forces and hydrophobic groups act together in order to minimize the disruptive effects on the hydrogen bonded water network; (3) hydrogen bonding, a unique type of dipole/dipole interaction; (4) Van der Waals forces, a very weak interaction between two very close atoms due to their fluctuating electrical charges.

Perhaps the most obvious means of protein binding to a surface is via electrostatic bond formation between charged groups on the protein and oppositely charged surface sites. The proteins in plasma are negatively charged at normal blood pH levels and thus would be expected to bind to cationic surfaces. The sensitivity of electrostatic bonding to the ionic composition is of some significance for the adsorption of proteins from blood. In this case, it is almost certain for protein bound in such a manner to experience adsorption which would be reversible and subject to continuous exchange with other ions in the blood. Protein adsorption appears generally to be irreversible or at best partially irreversible on many of the surfaces of interest for blood-contacting applications.

Another possible protein – surface interaction, which depends very strongly on surface chemical composition, is hydrogen bonding. This would be likely only for relatively polar substrates. H-bonding is an extremely important phenomenon in the internal structure of proteins, contributing strongly to the stability of the α-helical and β-sheet structures. Its role in binding to surfaces has not been thoroughly assessed.

Perhaps the most important general binding mechanism in the present context is that of hydrophobic interaction. The hydrophobic interaction may be defined as the interaction of nonpolar groups in aqueous media. Proteins are known to contain hydrophobic "patches" on their surface that would be available to interact with hydrophobic contacting surfaces. The heat change involved in such interactions is small, and the major thermodynamic driving force is believed to be the gain of entropy when structured water at the hydrophobic interface is transferred to the bulk solution.

2.5.3 Thermodynamics

Free energy change ΔG is defined as follows:

$$\Delta G = \Delta H - T\Delta S$$

where ΔH is the change in enthalpy, ΔS is the change in entropy, and T is temperature in K. For a process to be spontaneous, as is protein adsorption, ΔG of a system must be negative. Thus, either ΔH must be sufficiently negative or ΔS sufficiently positive to give a net negative result for ΔG that is, the adsorption must be either sufficiently exothermic or have a sufficient gain of entropy. Enthalpy changes can be obtained directly from calorimetric measurements. At the isoelectric point where proteins are effectively uncharged, adsorption is usually endothermic, and therefore the adsorption is driven by an increase in entropy.

It is clear the that exothermic characteristics can arise from the various interactions between proteins and surfaces. These interactions can, in principle, be covalent, electrostatic, or H-bonding as well as various nonbinding interactions. Entropy gaining mechanisms result mainly from the disruption of structured water either near the protein or near the surface, i. e. from hydrophobic interactions (Voet, 1995).

2.5.4 Transport Theory

In most situations of practical interest, where a solution of protein is flowing past a solid surface, the protein reaches the surface by a combined diffusion-convection process. Proteins bind to surfaces by various mechanisms. The binding process for any particular case is governed by its intrinsic kinetics. Transport and binding are sequential events of an adsorption process, so that the adsorption rate may be controlled by one or the other or both depending on their relative rates. If protein transportation is faster than binding, protein adsorption is dominated by binding. Vice versa, protein adsorption is controlled by protein transportation.

Well-developed mathematical models are available to describe protein adsorption in terms of convective-diffusion and binding processes. With rate data of sufficient accuracy, the controlling step can be determined based on fitting the data to models. When proteins reach a surface by diffusion along a concentration gradient, classical diffusion theory should be applicable. This would be subject to experimental verification in terms of adsorption measurements. Assuming adsorption is instantaneous upon protein arrival and protein concentration at the interface is zero, the rate of adsorption is given by Andrade and Hlady (1986):

$$\frac{\mathrm{d}n}{\mathrm{d}t} = C_o\left(\frac{D}{\pi t}\right)^{1/2}$$

where C_o is the bulk concentration, D is the diffusion coefficient and t is time. At solid surfaces, enough data of sufficient accuracy do not exist to test this equation. Some reports suggested that increasing the flow should lead to an increased adsorption rate since the diffusion coefficient D is effectively increased by bulk fluid movement. Most reports (Brash and Samak, 1978; Chan and Brash, 1979) suggested that the flow has a negligible effect on either adsorption rate or on the amount adsorbed at steady state. This led to the conclusion that transport is not the rate-limiting step for protein adsorption. Diffusion coefficients for proteins, while considerably greater than those of colloidally sized particles, are considerably smaller than those of "conventional" small molecules. Nonetheless the associated transport rates appear to be sufficiently rapid that they do not limit the overall adsorption process. With regard to transport in flowing blood, the phenomenon of red-cell-augmented diffusivities has been frequently discussed (Brash, 1981; Verkleij et al., 1998). This augmentation resulted from shear-induced lateral migrations and rotations of the RBC, giving effects analogous to eddy diffusion. These phenomena are highly significant for the transport of platelets, and several papers have reported platelet diffusivities greatly in excess of those predicted by Brownian motion.

In summary, proteins move to a surface by molecular diffusion, which may be supplemented by convective transport in flow situations. The available evidence suggests that the transport step would not be an overall limitation on adsorption.

2.5.5 Adsorption Theory

2.5.5.1 Langmuir Model

The adsorption of single molecules, primarily from the gas state, onto solid surfaces has been studied for nearly a century. Langmuir observed that for many systems, equilibrium adsorption followed a relationship which was later to be referred to as the Langmuir isotherm:

$$\rho = \frac{\rho_m}{1 + bc}$$

In this equation, ρ is the adsorbed surface concentration; c is the bulk solution concentration; ρ_m is the surface concentration corresponding to monolayer coverage; and b is a constant.

A typical Langmuir isotherm is shown in Fig. 2.7. The Langmuir isotherm assumes a maximum of monolayer adsorption and that only specific adsorption sites are available for adsorption. In addition, each adsorption site is assumed to have the same adsorption energy, and no lateral interactions are allowed between adsorbed molecules and the bulk phase.

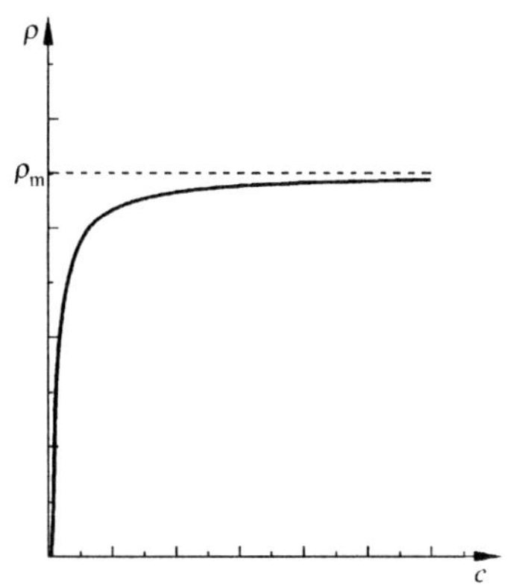

Figure 2.7 A typical Langmuir isotherm indicating the relation between protein adsorption and concentration

For many systems involving gases and solids, Langmuir's model adequately explains adsorption behavior. The Langmuir adsorption isotherm has also been employed to interpret the behavior of proteins adsorbing on solid surfaces, despite its shortcomings for protein adsorption. A major one is that the Langmuir model requires adsorption to be reversible while protein adsorption in most systems examined to date is irreversible on a realistic time scale. There seems little doubt that this stems from multivalent binding interactions which are typical of protein adsorption. However, the data in many experimental studies appear to fit the Langmuir equation well and some authors have interpreted this as confirmation that the Langmuir mechanism is applicable. Estimates of affinity constants for binding have been made from such data fits, although there is considerable skepticism about this practice. Such estimates should not be considered as thermodynamic equilibrium constants, and they can only be regarded as a qualitative indication of binding affinity.

2.5.5.2 Freundlich Isotherms

Some protein adsorption isotherms are not Langmuirian but instead fit a Freundlich isotherm,

$$\theta = kc^{1/n}.$$

In this equation, which is a power function; θ is the fractional surface coverage; k is a constant, c is the bulk phase protein concentration; and n is another constant, generally greater than 1.0. An example of the Freundlich isotherm is shown in Fig. 2.8.

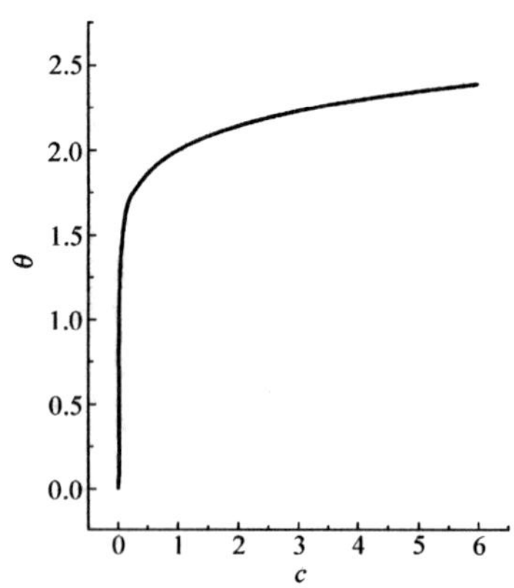

Figure 2.8 A typical curve of the Freundlich isotherm

The Freundlich isotherm has traditionally been explained in terms of heterogeneous adsorption processes. Adsorption from a solution containing only one type of protein is concentration dependent, rising toward a "saturation" amount at a higher concentration level which approximates the value expected for a close-packed monolayer. However, the saturation values are frequently above or below monolayer coverage by a factor of 2 or more, and upon close inspection, the data typically reveal that the saturation is never reached, but instead adsorption increases more slowly at higher concentrations than at lower concentrations. Nonetheless, the Freundlich monolayer model is frequently applied to protein adsorption isotherms. Deviations from the monolayer saturation values are ascribed to a lack of knowledge of true surface areas or to variations in packing due to orientation of the molecules (e. g., "side-on" versus "end-on" orientations). However, the Langmuir model of hard, nondeformable spheres adsorbing to a homogeneous surface without lateral interaction between the adsorbing molecules, originally devised for gas adsorption, is not a realistic model for protein adsorption.

2.5.6 Factors Influencing Protein Adsorption

2.5.6.1 Molecular Properties of Proteins

The surface activity of proteins is a fundamental property of these complex macromolecules that is generally recognized to be responsible for protein adsorption at the interface (Horbett and Brash, 1987). The size, charge, structure, and other chemical properties that influence surface activity are all related to their amino acid sequence, which is fixed for each type of protein, but varies greatly among proteins. Thus, differences in surface activity among proteins arise from variations in their primary structure.

Size is an important determinant of surface activity because proteins are thought to form multiple contact points when adsorbed to a surface. Larger molecules may have more contact points.

The charge distribution of proteins is likely to influence the surface activity because it is known that most of the charged amino acids reside on the exterior of protein molecules. These charged residues must therefore come into close proximity with the surface in the adsorption process. Experimentally, proteins have frequently been found to exhibit greater adsorption at or near the isoelectric pH, presumably because charge – charge repulsion among the adsorbed molecules is minimized under these conditions (Feng and Andrade, 1994). The role of the protein surface charge is especially important and probably predominant at interfaces with fixed ionic charges, as shown by the ability to adsorb proteins to ionized matrices. Adsorption to this type of surface is strongly affected by the degree of opposite charge on the protein and the degree of competition provided by like charged ions in the buffer. Adsorption to charged matrices is the basis for the widely applied separation of proteins by ion exchange chromatography.

Structural factors important in the surface activity of proteins are not well understood. Proteins that are likely to unfold to a greater degree or unfold more rapidly would be more surface-active. Thus, proteins with disulfide cross-linking would be less likely to unfold rapidly or completely and therefore be less surface-active.

Chemical differences among proteins arising from the particular balance of amino acid residues in each protein, probably, are also important factors influencing the surface activity of proteins. The amphipathic nature of proteins, due to the presence of hydrophobic, hydrophilic, and charged amino acid side chains, provides an opportunity of bonding to sites that vary considerably in chemical nature. Thus, for a particular surface, some proteins may have more of the type of residue that favors bonding to the kind of adsorption sites prevalent on this surface, and therefore would be more surface-active than other proteins. Furthermore, proteins that are more hydrophobic may be preferred on many surfaces, especially in view of the apparent importance of hydrophobic interactions. Lastly, the differences in solubility are important indicators of differences in surface activity.

2.5.6.2 Surface Properties

Surface properties have an enormous effect on the rate, the extent, and mechanisms of protein adsorption.

The broadest, most widely accepted generalization regarding surface properties concerns hydrophobicity and holds that the more hydrophobic the surface is, the greater the extent of protein adsorption is. The hydrophobicity "rule" is strongly supported by studies of hydrophobici gradient surfaces (Elwing et al., 1987).

Electrical charge is another general surface property that has been thoroughly investigated in relation to protein adsorption, although no clear consensus has been developed as to its effects. It seems obvious that attractive interactions should dominate between opposite charges and repulsive interactions between like charges. Although this is often the case, charge effects can be confounded by "lurking" factors such as small multivalent counterions bridging the gap between a protein and a surface having the same charge, which would normally be expected to repel each other. Also, proteins are large molecular entities which usually contain many charged groups, some negative and some positive. The question of whether they behave as particles with a net "point" charge or as assemblages of individual charges thus arises. Many research groups have observed maxima in adsorption capacity at the isoelectric point of the protein, suggesting a strong global charge effect.

Surface chemistry also has a great influence on protein adsorption. Considerable evidence has been accumulated showing that sulfonate-containing surfaces have high affinity for protein adsorption. Several reports have noted the extensive and strong adsorption of fibrinogen to sulfonated polyurethanes from a buffer and from plasma or blood. Thrombin behaves similarly on similar materials. Fibrinogen has also been shown to adsorb strongly to sulfonated silica surfaces, and sulfonated polystyrenes have been found to adsorb large amounts of proteins from plasma. It may be inferred that these interactions are nonspecific and indiscriminate with respect to protein type, i. e., all such surfaces appear to adsorb all proteins extensively and with high affinity. Over the past several years, many reports (Tani, 1996; Woodhouse et al., 1994) have appeared in which surfaces have been designed for the selective adsorption of specific proteins. Affinity chromatography matrices provide many examples of selected materials for the binding of specific proteins.

2.6 Surface Modifications for Enhanced Biocompatibility

2.6.1 Introduction to Surface Modification

Most conventional materials do not meet the criteria for serving as biomaterials. An effective approach for developing a clinically applicable biomaterial is to modify the surface of materials that already have excellent bulk properties. Surface modification is an effective means to alter biological interactions of a particular material and offers a number of advantages in biomedical applications. By changing only the composition at the outermost surface of a biomaterial, its mechanical properties and fabrication methods of an implant can remain unaffected. Since surface modification techniques require only minute amounts of material to completely alter the surface properties of a device or implant, expensive coatings can be economically used in commercially competitive biomedical products. In addition, surface modification can provide accessible and chemical functional groups that are used to immobilize drugs, enzymes, antibodies, or other biologically active species for a variety of biomedical applications.

Depending upon the medical applications, surfaces are mainly modified either by passivation to prevent biofouling or by activation to incorporate a specific functionality for binding cells and tissues. The past decades have witnessed considerable activity in searching for nonfouling surfaces. As discussed in previous sections, proteins are known to play a critical role in the process of cell – biomaterial interactions. When the materials contact blood or other body fluids, a layer of protein molecules always first adsorbs on the surfaces of the materials. The adsorbed protein molecules then mediate the interactions between various cells and the material surfaces. Attachment of cells on material surfaces can occur in the absence of any adsorbed proteins, but cells, in general, fail to spread under this condition and this subsequently results in cell detachment and death (Horbett et al., 1996). A decrease in protein adsorption often leads to a reduction in cell adhesion, including blood and tissue cells. Therefore, protein-resistant surfaces can be generally referred to as non-fouling surfaces. Among all biocompatible polymers, polyethylene glycol (PEG) is one of the most effective nonfouling polymers (Harris, 1992; Zalipsky, 1995; Amiji and Park, 1993;

McPherson et al., 1995). Since PEG coatings play a crucial role in improving material's biocompatibility, PEG properties, protein-resistant mechanisms, and PEG surface modification on materials used for BioMEMs applications will be discussed in detail in the following subsections.

2.6.2 Surface Characterization

It is well recognized that understanding the nature of a biomaterial surface is essential both for understanding the interactions between materials and living systems and for fabrication of biomaterials and biomedical devices. Surface analysis methods provide the means to gain this understanding. Many methods are now available for surface characterization, with each providing a unique piece of information about the nature of a surface. Several commonly used surface characterization approaches are described here.

A contact angle goniometer is used to measure the wettability of surfaces. This approach offers a number of unique advantages over other surface characterization techniques. First, contact angles are sensitive only to the outermost atomic layers responsible for surface energetics that drive protein adsorption. The second advantage is that the results of contact angle measurements can be interpreted directly in terms of the surface energetics of the adsorbed layer. The third is that contact angle measurements can be used to confirm the immobilization of biofilms on substrates based on the change in the wettability of substrates. Fourth, it is inexpensive and easy to perform. Contact angle measurements provide insight into how the surface will interact with the external world.

Ellipsometry is commonly used to measure the thickness of biofilms and proteins immobilized on substrates. The technique has been known for almost a century and has wide used for many applications. It has been used in semiconductor research and production to determine the properties of thin films at interfaces. Over the past decades, ellipsometry has become more interesting to researchers in other disciplines such as biology and medicine. As a matter of fact, ellipsometry has become a standard technique for the analysis of protein adsorption, especially suitable for silicon, with its smooth and reflecting surfaces (Wannerberger and Arnebrant, 1996; Mandenius, et al., 1984; Malmsten and Lassen, 1995; Walivaara et al., 1995). Ellipsometry is a simple and quick analytical technique. In addition, ellipsometry adds considerably to the understanding of protein adsorption, particularly with regard to the thickness of the protein layer and hence to protein conformation and to the type of packing. The amount of absorbed protein in ng/cm^2 is usually estimated from thickness data using the method proposed by Stenberg and Nygren (1983).

Electron spectroscopy for chemical analysis (ESCA), also called X-ray photoelectron spectroscopy (XPS), is used to analyze the surface composition of materials. XPS is based upon the emission of electrons from matter in response to the irradiation by photons of sufficient energy to cause the ionization of core-level electrons. These electrons are emitted with characteristic energies of the atoms, which in turn provides information on the electron structure of the material under investigation. Since photoemitted electrons have little ability to penetrate matter, only those electrons emitted near the surface can escape and be counted. Therefore, XPS is a surface-sensitive technique with information on typical penetration depths of 1 – 7 nm, corresponding to 4 – 20 atomic or molecular monolayers. The penetration depth is determined by the mean free path of electrons. This technique is especially valuable for analyzing polymers and other organic materials.

The scanning electron microscope (SEM) is widely used to obtain information on surface morphology. It is the first analytical instrument used when a "quick look" at a material is required and the light microscope no longer provides adequate resolution. SEM images of surfaces have great resolution and depth of field with a three-dimensional quality that offers a visual perspective. The SEM is an important corroborative method to use in conjunction with the above mentioned surface analytical methods. Nonconductive materials observed in the SEM are typically coated with a thin, electrically grounded layer of metal to minimize negative charge accumulation from the electron beam. If the metal coating is truly conformal, a good representation of the surface geometry will be obtained.

Surface plasmon resonance (SPR) is an optical technique that is widely used as a valuable tool and is rapidly gaining recognition for biomaterial characterization (Green et al., 2000). SPR is capable of monitoring any dynamic process, such as adsorption or degradation, to a wide range of biomedically relevant interfaces in real time without labeling or complex sample preparations. It can rapidly obtain information on the rate and amount of films and proteins, as well as their dielectric properties and the association/dissociation kinetics of molecules adsorbed on the surface. Many surface analytical techniques have been applied to the study of biomaterials, but few are able to monitor dynamic interactions within a fluid environment. SPR may be tailored to model similar conditions encountered *in vivo*.

2.6.3 Applications of PEG Surface Modification

PEG surface modification can be used for a wide range of biomedical microdevices for the improvement of material biocompatibility and device efficacy. A few examples will be discussed here.

2.6.3.1 Biosensors

Exciting developments have occurred over the past several decades in biosensors (Lee et al., 2000; Schugerl et al., 1996; Marrazza et al., 1999; Collings and Caruso, 1997). Bioreceptor molecules (enzymes, membrane proteins, and antibodies) have been immobilized on transducer surfaces to develop sensors for agricultural, biological, environmental, defense, and food processing applications (Mulchandani et al., 1999; Bucher et al., 1999; Hengerer et al., 2000; Samsonova et al., 1999). Medical and clinical applications are the most lucrative and important avenues for biosensors and, accordingly, much research and development have been devoted to them. The trend toward home health care and self-monitoring, such as daily glucose testing by diabetics, represents the "holy grail" for medical biosensing. Although still in its infancy, the biosensor industry holds great promise in many aspects of health care, from home care to affordable diagnostics, and to sophisticated and critical health care.

A sensor is a device that generates a signal in response to a physical stimulus, which can be used for measurement, interpretation, and control (Tolles, 1996). A sensor is usualy comprised of three essential components: a detector recognizing the physical stimulus, a transducer converting the stimulus to a useful, invariably electronic output, and an output system involving amplification and display. The term biosensor is now generally applied to any device which incorporates a biological material, such as tissues, microorganisms, organelles, cell receptors, enzymes, antibodies, nucleic acids, etc., with a physiochemical transducer. Biosensors usually yield an electronic signal which is proportional to the concentration of a specific analyte or group of analytes. Sensors with molecules or cell receptors are called molecularsensors and cell sensors, respectively.

The immobilization of biomolecules to the transducer surface is one of the key elements in the construction of a biosensor. However, the development of current sensor technology is hindered by the adsorption of nonspecific biomolecules onto surfaces and a loss of the native conformation of biomolecules immobilized at solid substrates (Williams and Blanch, 1994; Flounders et al., 1995; Ramsden, 1996). Many research groups reported that PEG exhibited favorable properties when used as a tether or to link the couple of active biomolecules to a surface (Holmberg et al., 1992). The results demonstrated that the bound PEG reduced the adsorption of undesirable molecules and did not cause the denature of bounded active molecules. It is, therefore, of great interest to modify material surfaces with PEG films that would inhibit nonspecific protein adsorption on the sensor surface and prevent the denaturation of functional biomolecules immobilized on the sensor surface.

2.6.3.2 Nanofilters

Micromachining precision has been advantageously employed for manufacturing silicon membranes for the size-based separation of biomolecules and biologicals (Ferrari et al., 1996; Tu et al., 1998; Kittilsland and Stemme, 1990; Brody et al., 1995; Woolley and Mathies, 1994). In view of the capability to produce uniform pores of desirable dimensions by microfabrication technology, a microfabricated silicon membrane is especially suitable for biofluid purification, such as viral elimination. These biofilters are designed to remove known or unknown viruses from blood products. Human blood, proteins derived from blood and blood plasma have become indispensable therapeutic agents for combating many life-threatening diseases. Unfortunately, these products have in the past also transmitted viral diseases. The risk of the transmission of the major transfusion-transmissible viruses has been dramatically reduced over the past decade as a result of improved donor selection, historical screening, virus discovery, and the implementation of progressively more sensitive serological assays. Nonetheless, viral transmissions still occur, resulting in continued public anxiety (Morgenthaler, 1989, 1994).

Although many techniques exist for inactivating or removing viral species from biological fluids, there are limitations or disadvantages associated with each method (Wyatt et al., 1993). Physical methods which involve pH adjustments, ultracentrifugation, or temperature treatment are not only difficult to perform on a large scale but also have ramifications for the integrity of the final product. Pasteurization is an effective method but only for enveloped viruses. Solvent/detergent treatment is also used as a method of choice for the inactivation of enveloped viruses. The addition of chemicals creates concerns due to the carcinogenicity of the chemicals and yet, have not been proven effective for complete inactivation. There are no identified general methods whereby chemicals will only kill all viruses without damaging blood cells. Treatment of biological materials with UV- or gamma-irradiation has also been shown to be disruptive to the functionalities of the products to be treated and yet quite costly.

Polymeric retention filters have been used for virus removal in certain processes (Wyatt et al., 1993; Morgenthaler, 1994). However, the fabrication processes associated with polymers are difficult to control, and generally are incapable of yielding accurate and repeatable pore sizes, particularly in the nanometer range. Therefore, there is a great deal of interest in searching for more user-friendly methods that maintain product integrity and ensure the absence of viral agents with reproducible assurance.

Advanced microfabrication technology offers unparalleled advantages for manufacturing silicon-based membranes with high precision in pore sizes, superior mechanical strength, and extraordinary thermochemical inertness. Such silicon nanofilters have been developed by Ferrari and colleagues (Ferrari et al., 1996). However, the problem of biofouling associated with protein adsorption remains to be solved if the nanofilters are to be of any clinical use. In addition, like all blood-contacting materials, the use of nanofilters for viral elimination is also limited by surface-induced thrombosis initiated by the adsorption of plasma proteins, followed by the adhesion of platelets. The biocompatibility and efficacy of silicon nanofilters may be enhanced by surface modification with biocompatible PEG films, which would reduce protein adsorption and thrombosis, preventing clogging, and sustaining the functionality of the filter.

2.6.3.3 Silicon Biocapsules

Polymeric biocapsules emerged over thirty years ago as a possible way by which transplanted cells could be protected from immune rejection by enclosure within a semipermeable membrane (Lanza and Chick, 1995; Zondervan et al., 1992). The transplantation of pancreatic islets of Langerhans has been reported to be a potential therapy for Type I Diabetes, the third-largest cause of death in the USA. Pancreatic islets normally secrete insulin that enables other cells to take up the sugar glucose from the blood for energy.

Daily injections of insulin extracted from animals are generally used as a tentative way to cure the disease. Unfortunately, many patients eventually suffer from potentially devastating diabetes-related disorders. Microscopic blood vessels can slowly become damaged, often resulting in blindness, or kidney failure, etc. An ideal treatment would be transplantation of islets that would provide the proper insulin production. A major obstacle to transplantation is the limited supply of donor human organ or islets. The use of xenografts, or cross-species transplants offers great promise by employing nonhuman tissue as an essentially unlimited donor resource. Successful clinical trials for diabetic patients, however, are hindered by graft rejection by the immune system of the human body. An attractive solution to this problem is the immunoisolation of islets by microencapsulation. This technique utilizes artificial membranes to isolate cell grafts from the host immune system, while preserving the viability of the transplanted cells.

Recently, silicon-based microfabricated biocapsules have been introduced as a potential therapy for diabetes. Experiments have shown that these silicon biocapsules maintained viability and functionality of many cells (Ferrari et al., 1996). It has also been demonstrated that pancreatic islets functioned physiologically both *in vitro* and *in vivo* by secreting therapeutic molecules as long as they were sufficiently immunoprotected (Desai et al., 1998).

Membrane fouling associated with protein adsorption, fibrosis, and biocompatibility are the major problems to be solved to ensure the success of silicon biocapsules for cell transplantation. Tight control over protein adsorption is needed, not only to ensure long-term potency of the pores in the immunoisolation membrane, but also to minimize fibrotic response. Biocapsules are encapsulated by fibrous tissue upon implantation when they are in contact with tissue. Excessive fibrosis blocks the entrance of nutrients to cells within the capsule, leading to the death of the transplanted cells and the failure of the biocapsule device. A decrease in protein adsorption usually leads to a decrease in cellular attachment. Thus, it is necessary to develop silicon surfaces with reduced protein and cell adherence for the prevention of fouling and fibrotic responses.

The therapeutic effectiveness of silicon biocapsules may also be undermined by the attack of the patients' antibodies. The possibility of reduction of the host's immune response to the biocapsule poses a great challenge to modern immunology in relation to the development of diabetic therapy. PEG-modified surfaces are well known to reduce immunogenicity and antigenicity (Harris, 1992). It is of considerable interest to modify silicon biocapsules with PEG films to reduce the adsorption of antibodies and explore potential advances in silicon-based, novel, diabetic therapy.

2.6.4 Characteristics of PEG Coatings

PEGs are viscous, colorless liquids when their molecular weights are less than 1000. PEGs with higher molecular weights are waxy, white solids. The melting point of PEGs is determined by the molecular weight, approaching a plateau at about 67℃. PEGs possess a variety of properties pertinent to biomedical and biotechnical applications (Harris, 1992; Bailey and Koleske, 1976).

PEGs are soluble in water, toluene, methylene chloride, and many organic solvents; insoluble in ethyl ether, hexane, ethylene glycol; are highly mobile and have large exclusion volumes in water; form two-phase systems with aqueous solutions of other polymers; nontoxic; FDA approved for internal consumption; biocompatible With biological materials; and weakly immunogenic. Covalently linked PEG will solubilize other molecules and render proteins nonimmunogenic and surfaces protein-rejecting.

One possible mechanism of PEG's passivity might be its minute interfacial free energy. PEG – water interfaces have very low interfacial free energies and thus low driving forces for protein adsorption. Proteins at or near a low energy interface would not experience any greater effects from the interface than they do from the bulk solution. However, other neutral, hydrophilic polymers, such as agarose, dextran, methylcellulose, polyacrylamide, etc., also have very low interfacial free energies, but they appeared to be more interactive than PEG surfaces. This implies that other factors are involved in PEG's passivity.

The inert character of PEG surfaces is believed to be due to the solution properties of the polymer, its molecular conformation in an aqueous solution, and the fact that it is completely noncharged. The oxyethylene unit, $-OCH_2CH_2-$, has a C – O – C bond angle of 110°, not far from the H – O – H angle. The structural similarity to water and strong hydrogen bonding to either oxygen atoms make it completely miscible with water in all proportions. In pure liquid water, hydrogen bonding would result in a highly connected network of tetrahedrally coordinated water molecules. It was suggested that PEG segments nicely fill out the voids in the water structure and minimally perturb the structure of water itself, thereby minimizing the tendency for hydrophobic interactions (Amiji et al., 1993). The hydrophilicity and unique solubility properties of PEG can produce surfaces that are in a liquid-like state Where the polymer chains exhibit considerable flexibility and mobility. The rapid movement of hydrated PEG chains attached to a surface can influence the microthermodynamics at the protein solution/surface interface and prevent adsorption or adhesion of proteins. Moreover, the rapid motion of the molecule gives an approaching protein little time in which to form a positive interaction.

Steric stabilization could be another factor that enables PEGs to prevent proteins and other biomolecules from approaching a surface (Amiji and Park, 1993). Steric stabilization usually has two contributions, an elastic term and an osmotic term. The elastic term results from the loss of conformational entropy when two surfaces approach each other, causing a reduction in the available volume for each polymer segment. Thus, when a protein approaches the PEG-modified surface, a repulsive force is developed due to the loss of conformational freedom of the PEG chains. The osmotic interactions may arise from an increase in the polymer concentrations on compressing the two surfaces. When a protein or another large molecule in water solution approaches the surface, the number of available conformations of PEG segments is reduced due to the compression or interpenetration of polymer chains by the protein, and an osmotic repulsive force is developed. Whether the protein would interpenetrate or compress the chains of the PEG, or both occur, depends on the density of PEG chains. If the PEG chains are dense, the compression is likely to dominate, otherwise, the interpenetration dominates.

2.6.5 PEG Surface Modification

PEG has been studied extensively as one of the most effective biocompatible coatings due to its unique structural properties. For the past decades, the science and technology of PEG immobilization on polymeric surfaces have recorded major advances (Morpurgo et al., 1996; Prime et al., 1993; Li et al., 1996). Commonly used surface modification methods for coupling PEG films on polymers include radiation grafting, plasma grafting, and physical adsorption of copolymers of high molecular weight.

The PEG coatings deposited by the methods cited have proven quite successful in depressing protein adsorption and cell attachment. However, the thick coatings produced by these methods limit their viability in applications where nanometer-scale dimensional requirements must be enforced, such as in the field of BioMEMS. Other concerns about these techniques are the complexity, high equipment costs, and the difficulty in control of coating uniformity which might eventually affect the materials blood compatibility. On the other hand, the self-assembly technique has generated increasing interest in recent years due to its simplicity and capability in immobilizing organic moieties onto material surfaces with molecular-level control over the surface properties. Films with different functionalities have been successfully immobilized on different material surfaces. However, few attempts have been made to date to immobilize PEG films MEMS materials using these self-assembly techniques for improved biocompatibility. This section will review recent developments on the surface modification of materials (oxides and metals) with thin PEG coatings. Immobilization of PEG films by self-assembly techniques is the focus of this section they are simple, quick, cost-effective, and applicable for BioMEMS.

1. Physical Adsorption of PEG on Silicon by Self-Assembly

PEG films can be physically adsorbed on silicon. Zhang and Ferrari (1997) reported that protein adsorption on silicon could be tailored by preadsorbing PEG or other surfactant molecules (Tween 20) on silicon surfaces with different hydrophobicity. In their study, the authors examined the effects of immobilized PEG films on the resistance to protein adsorption. Figure 2.9 shows, schematically, proteins immobilized on three types of silicon surfaces: untreated silicon, HF treated silicon, and methylated silicon with water contact angles <10℃, ~60℃, and ~100℃, respectively. The influence of Tween 20 on the reduction of protein adsorption on silicon surfaces with differing wettabilities was investigated and is shown in Fig. 2.10. Protein adsorption was found to depend on the degree of hydrophobicity of the surfaces. The protein adsorption study showed that the reduction protein of adsorption by the surfactant was most pronounced on the most hydrophobic surfaces. While a reduction of 90% in protein adsorption on methylated surfaces by PEG surfactant was achieved, only a 15% reduction on hydrophilic surfaces was obtained.

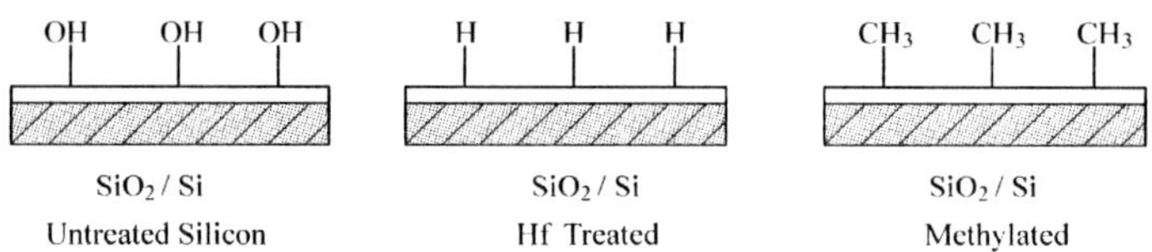

Figure 2.9 A diagram showing silicon surfaces modified with various chemical functional groups

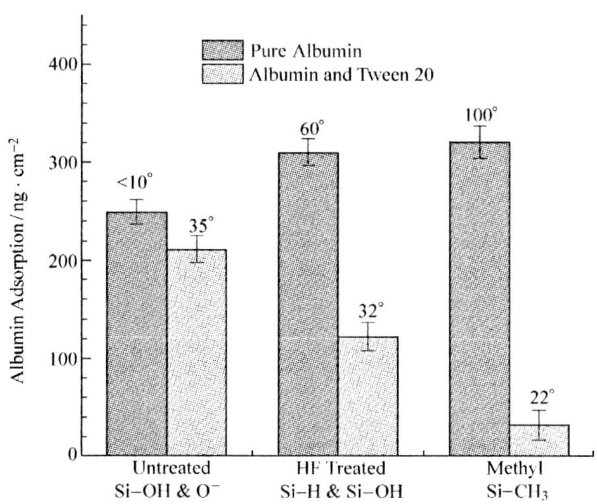

Figure 2.10 Albumin adsorption on various surfaces from a pure protein and a protein-surfactant mixture

2. Covalent Immobilization of PEG by Self-Assembly

Self-assembled monolayers (SAMs) are molecular assemblies that are spontaneously formed on a substrate upon exposure to an active surfactant solution (Ulman, 1991; Whitesides, 1995). Two families of self-assembled molecules have received the most attention. The first is based on silanization, the reaction of trichloro- or trialkoxysilanes with a hydroxylated material surface, and the second is based on the strong sulfur – gold interactions when gold surfaces are exposed to a thiol and disulfide solution. Much work has been performed in the last decade that involved chemical modification of surfaces with such monolayers. Here, we focus on the general approaches in modifying gold and oxide surfaces with self-assembly PEG films.

(1) PEG on silicon

Silanization is the most widely used method to modify a variety of surfaces including glass, silicon, and metals (Burns et al., 1995; Balachanderz and Sukenik, 1990). Figure 2. 11 shows an example of a reaction scheme for immobilizing PEG film on silicon by self-assembly (Zhang and Ferrari, 1998a). In this approach, surface modification mainly involves two steps, PEG conjugate synthesis and formation of PEG self-assembled monolayers. PEGs are covalently coupled on silicon surfaces by self-assembly through the functionalization of a PEG precursor by the formation of $SiCl_3$ groups at its chain ends, and followed by the reaction of the surfaces with compounds of PEG-$OSiCl_3$. During the self-assembly process, Si – Cl bonds react with both OH groups and the trace water adsorbed on the surface to form a network of Si – O – Si bonds (Zhang and Ferrari, 1998b). A monolayer in which the molecules are connected, both to each other and to the surface by strong chemical bonds, will be self-assembled on the surface. A monolayer or multilayer of PEG will be then formed if the starting

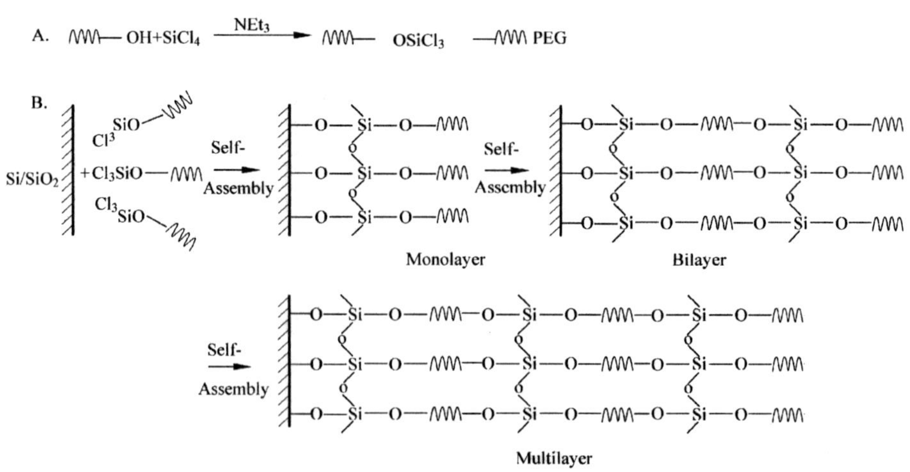

Figure 2. 11 Schematic representation of the immobilization of PEG on silicon surfaces. A. synthesis of PEG silane derivative; B. Formation of self-assembled PEG multilayers onto silicon surfaces

PEG material has a $-OCH_3$ or $-OH$ at one terminal, respectively. The detailed reaction procedure can be found in Zhang and Ferrari (1998c). PEG modified silicon surfaces show an effective reduction on protein adsorption and cell adhesion.

Figure 2.12 shows the time dependence of albumin, fibrinogen, and IgG adsorption on unmodified and modified silicon surfaces. The reduction of albumin, fibrinogen, and IgG adsorption on PEG modified surfaces at 120 minutes was about 75%, 65%, and 80% (Zhang and Ferrari, 1998a). Figure 2.13 shows the adhesion of fibroblast cells to PEG treated and untreated silicon

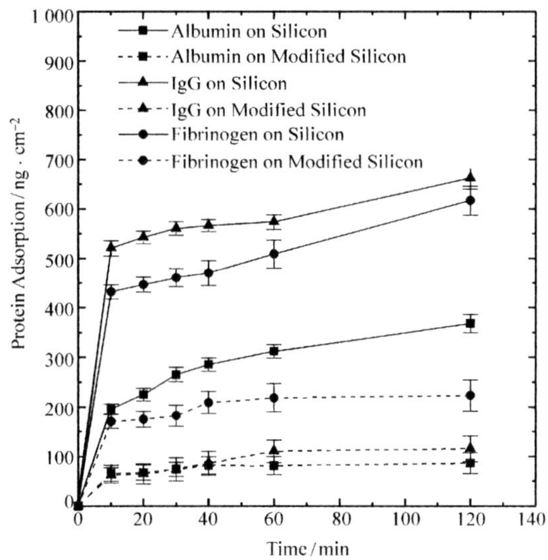

Figure 2.12 Adsorption of albumin, fibrinogen, and IgG onto silicon and PEG modified silicon surfaces

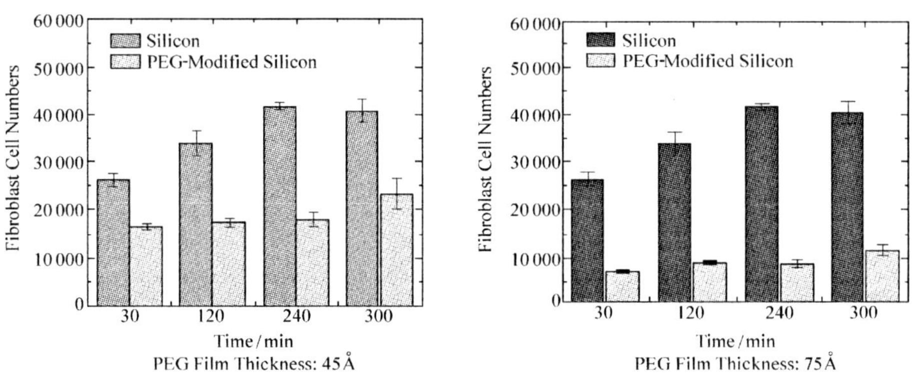

Figure 2.13 Fibroblast adhesion on unmodified and modified silicon surfaces with PEG films of different thickness

substrates with respect to time (Zhang and Ferrari, 1998a). The reductions in the adhesion and proliferation of fibroblast cells increased with time and the thickness of PEG films. The maximum values were about 55% - 62% and 70% - 80% for the PEG films of 45 Å and 75 Å in thickness, respectively.

(2) PEG on gold

The immobilization of PEG films on gold can be achieved by grafting thiol groups on one terminus of PEG oligomers and followed by reacting gold surfaces with a mercaptan PEG oligomer in an organic solution. An example of a simplified reaction scheme is shown in Fig. 2.14. A large amount of work on the synthesis of PEG thiolates and the immobilization of PEG on gold substrates for protein resistance was performed by Whitesides and co-workers (Pale-Grosdemange et al., 1991; Prime and Whitesides, 1993; Mrksich, 1996). PEG-modified surfaces showed effective protein resistance.

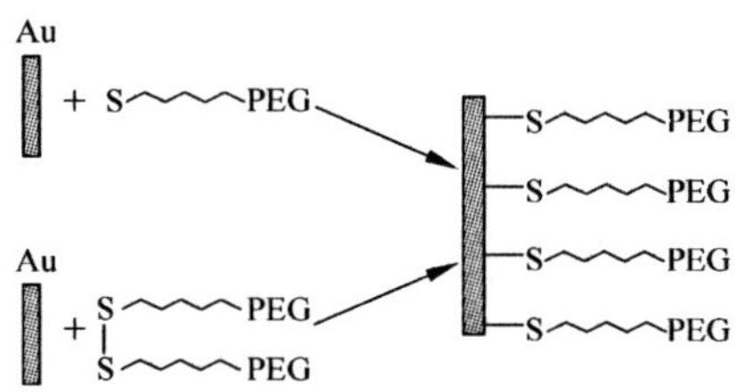

Figure 2.14 Formation of gold-thiol monolayers

(3) PEG patterned surfaces for controlled protein adsorption and cell adhesion

The ability to pattern proteins and cells on surfaces is crucial to the development of molecular, or cell-based sensors, and other miniature biological-electronic devices. The most common methods for patterning biomolecules are microcontact printing (Lahiri et al., 1999; Huang et al., 1994) and photolithography (Lom et al., 1993; Groves et al., 1997).

In the example shown in Fig 2.15, the microcontacting printing technique is used to create specific patterns of self-assembled monolayers of alkanethiols and to create areas with defined shape and size that support or resist protein adsorption for controlled cell adhesion. This is accomplished with a stamp made by casting polydimethylsiloxane into the desired pattern and "inking" the stamp with a thiol-terminated alkane. When the stamp contacts a gold surface, the alkanethiol is self-assembled within the stamp pattern onto the gold. The surface is then incubated with a second thiolate that would self-assemble onto the rest of the clean gold surfaces left after stamping. This results in a pattern of the first and second thiolates. The thiolates are bifunctional with one end group of the molecular chain binding to the surface and one end group remaining free to interact at the interface. In general, one thiolate is a thiol alkane which promotes protein adsorption. If adhesion proteins are used, cells can organize on the proteins. The other thiolate is a PEG-terminated alkanethiol that inhibits protein adsorption and cell attachment. A schematic diagram of protein and cell patterning is shown in Fig. 2. 15. Micro-contacting printing is a simple, low cost, and effective patterning method, but is limited to patterning one protein on a gold substrate.

The photolithographic technique can be used to create protein patterns on silicon/glass/metal substrates (Blawas and Reichert, 1998). The patterning is accomplished using chemical linkers to create a heterogeneous monolayer. Silane coupling agents and alkanethiols have been used for attaching proteins to silica, metal and gold surfaces, respectively. They form a robust bond with the substrate and protein, and withstand the harsh solvent systems required to remove a photoresist. A conventional photoresist technique applied to the patterning of proteins and cells is shown in Fig. 2. 16. The substrate is first coated with photoresist. Ultraviolet (UV) radiation decomposes the photoresist and renders substrates with some open defined areas. A silane for promoting protein adhesion reacts with the patterned substrate and binds within the open areas. The substrate

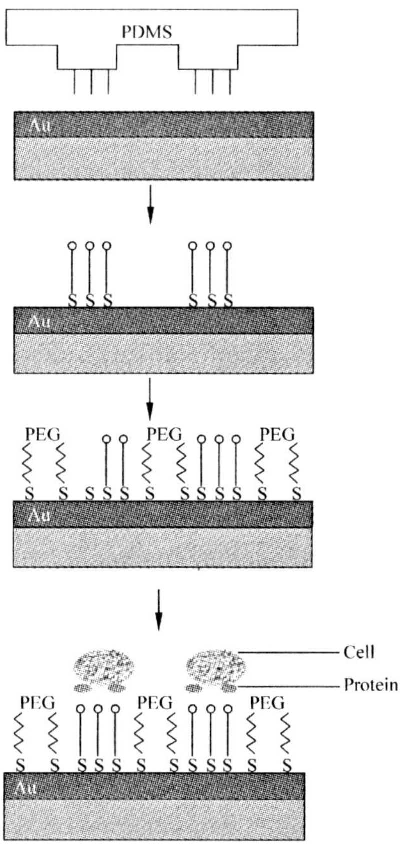

Figure 2. 15 Microcontact printing with a polydimethylsiloxane stamp with two thiols for protein and cell patterning

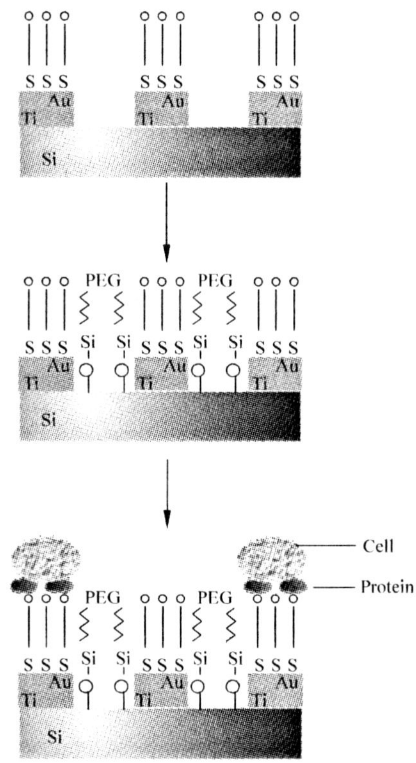

Figure 2. 16 Protein and cell patterning by photolithography

is then sonicated with acetone to remove the remaining photoresist. Thus, PEG-terminated derivatives are exposed to the rest of the substrate. There are many methods to pattern proteins using photolithography. One example is of that the method Tender et al. who used to create gold patterns by photolithography (Tender et al., 1996). In this method, alkanethiols were adsorbed to the entire surface and desorbed from the gold areas via voltage cycling. The surface was then incubated with hexadecanethiol that adsorbed to the clean gold surface. Protein preferentially adsorbed on the hexanedecanethiol and created a protein pattern.

There are several advantages of photoresist-based protein patterning. It allows proteins to be patterend on different substrate materials and to choose "background" silane where proteins are not bound. Its patterns can also be precisely controlled. The major drawback of photolithographic patterning is that the chemicals involved in the patterning processes, such as photoresists and residual solvents, would denature proteins and reduce their activity (Blawas and Reichert, 1998). Protein activity and confirmation are critical factors in determining sensor efficiency and subsequent cell adhesion for tissue engineering.

A new approach to covalent immobilization of proteins on patterned surfaces has been recently introduced through combined photolithography and chemical selectivity techniques (Veiseh et al., 2000). The pattern is created by the different affinity of thiolates and silane coupling agents toward gold and silicon surfaces, respectively. In the patterning process, protein immobilization is the last step, and no photoresist removal is involved. Protein conformation and activity are therefore retained. In this method, the surfaces are prepatterned with gold squares distributed regularly on a silicon surface using photolithography. The fabrication procedure with a simple liftoff microfabrication process is outlined in Fig. 2. 17. The chemical reaction scheme is shown in Fig. 2. 18. The

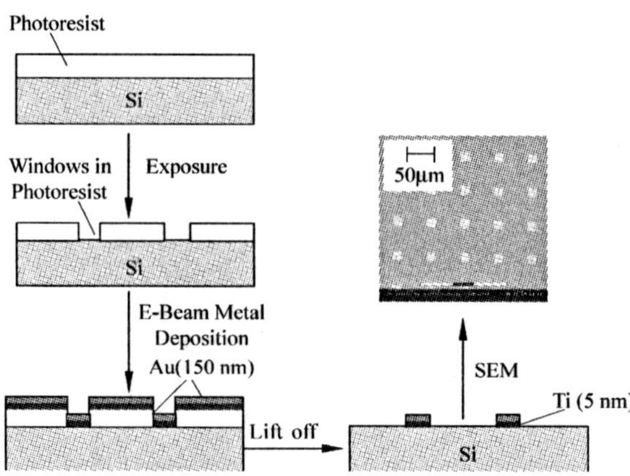

Figure 2. 17 Schematic illustration of fabrication procedure for generating gold patterns on silicon and a scanning micrograph of the prepatterned Au/Si surface

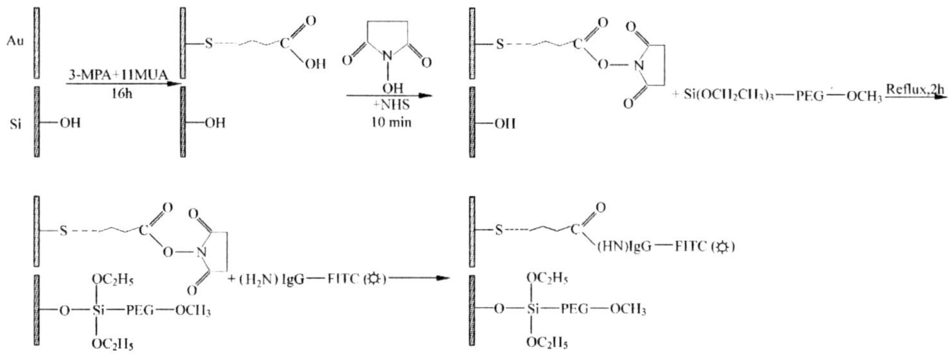

Figure 2.18 The scheme of chemical reactions for coating a patterned surface with protein on an Au surface and PEG on a Si surface

gold regions on the substrate are first immobilized with bifunctional solf-assembled monolayers (SAM) layers, followed by covalent immobilization of adhesion proteins which allow cell attachment against PEG backgrounds. The silicon regions of the substrate are modified with polyethylene glycol (PEG) to resist protein adsorption and cell adhesion.

In this patterning process, react pre-patterned gold – silicon substrates are reacted sequentially with ①carboxylate-terminated thiolates for the immobilization of SAM on gold, ②*N*-Hydroxysuccinimide/ 1-Ethyl-3-(3-dimethylamino-propyl) carbodiimide (NHS/EDC) solution for the activation of the carboxylates, ③the PEG – silane solution for immobilization of PEG on silicon, and ④proteins for immobilization on activated carboxylates. In this process, the attachment of thiolates and the reaction of the SAM with NHS/EDAC solution have no influence on PEG immobilization on silicon.

Similarly, the PEG silanization reaction has no effect on protein immobilization on the NHS activated carboxylates. This technique demonstrated excellent immobilization selectivity, and the PEG-modified silicon surfaces showed effective protein reduction.

2.7 Future Directions

The host/biomaterial interactions which follow implantation of any prosthesis or device involve a complex, but poorly defined series of events. A clear understanding of the mechanisms involved in the interactions at the cell – material interface is a prerequisite for the rational design of medical devices with optimal biocompatibility. The past two decades have witnessed a revolution in understanding molecular cell biology and the chemical processes of living organisms. It is clear that progress and advances in application of molecular cell biology to biocompatibility evaluation would lead to the development of improved biomaterials and have an impact on materials, biotechnology, medicine, and many other areas. There are many opportunities and challenges both in biocompatibility assessment and surface modifications for improved biocompatibility. In this section, some examples will be given to illustrate the future trends in these areas.

2.7.1 Biocompatibility Assessment at the Molecular Level

2.7.1.1 Assessment of Host Responses to Implanted Materials

The conventional biocompatibility assessment of host responses to implanted materials has been largely conducted at the cellular level. These techniques have been used to investigate cellular changes induced by materials, including cell attachment, migration, growth, differentiation, and apoptosis. However, they are incapable of revealing the mechanisms that govern these changes. Understanding the host response to implanted materials at the molecular level, on the other hand, may assist in elucidating the complex host/materials interactions at the cellular level. For this purpose, a molecular biological approach is used to study enzyme production, integrin receptor expression, chemokinesis, complement and metabolic responses, and the expressed messenger RNA induced by materials in contact with cells (Chou et al., 1998; Curtis, 1997; Curtis et al., 1998). In general, each specific material - cell interaction will induce characteristic biocompatibility indicators. For example, in bone tissue engineering, the activity of alkaline phosphatase (ALP), a group of membrane-bound phosphomonoesterases, can be used to determine the biocompatibility of scaffolds. Alkaline phosphate is one of the most widely used biochemical marker of osteoblasts and its activity correlates well with the osteoblastic function. Antibodies have unique ability to bind to their corresponding antigens with high specificity and affinity, and can be used to identify and locate the specific proteins and cytokines. Antibody technologies are particularly valuable in assisting the molecular biological approaches in effectively characterizing newly discovered cytokines, receptors or cell adhesion molecules. The combination of molecular and antibody technologies is perhaps a promising approach to meet the current challenges in biocompatibility assessment.

2.7.1.2 Effect of Surface Topography

It has been recognized that the surface topography of an implant is an important factor affecting cellular functions, including cell shape, migration, adhesion, and tissue organization. Many experiments demonstrated that cell shape affects gene expression (Curtis and Wilkinson, 1998; Ito, 1999). Cells not only look different on the surfaces of differing topography, but also function differently. Thus, the information on how the cell topography affects cell functionality at the molecular level may help provide a deep understanding of the nature and fundamental principles of cellular changes induced by surface topography. Again, the molecular techniques combined with antibody techniques can be a good approach to this application. For example, the molecular techniques could be used to detect changes in messenger RNA at the tissue – implant interfaces that are conjugated with the antibody tagged with gold or fluorescence. Information on localization and identification of characteristic biological macromolecules of cells induced by surface topography can be obtained with electron/light and fluorescence microscopy. These specific approaches to monitoring the outcomes of altering surface topography, combined with micromachining and micropatterning that can produce surfaces with desired topography, have the potential to provide the information that will enable future implants to be engineered with desired cell responses.

2.7.1.3 *In Vitro* Screening Methods for Evaluating Genotoxicity

Screening tests for assessment of material-induced mutagenesis *in vitro* are highly desirable and have extensively studied for the past three decades, in the hope of replacing cost – intensive and time-consuming animal experimentation. The Ames test, which involves the detection of mutations in metabolic functions of bacteria, is most widely used but is generally inapplicable to the assessment of the mutagenicity of metal alloys or microdevices. Many short-term tests of the material mutagenicity on mammalian cells in culture, such as chromosome aberrations and sister chromosome exchanges on the basis of the detection of end-point genetic changes, have been developed (Kilbey et al., 1977). However, these techniques relied upon prolonged and labor-intensive work, which involves isolation of nucleic acids, staining, immunoassay and/or radioactive assay, and microscopic examination of chromosomes.

A new method was recently introduced to screen the genotoxicity of metal ions and environmental carcinogens using synchrotron Fourier transform infrared microspectroscopy (Zhang et al., 1999; Holman et al., 1999). This technique allows for the noninvasive study of individual cells *in situ* including monitoring of chemical reactions occurring at a material – cell surfaces and of enzymatic reactions, as cells respond to materials and toxins. This technique is also capable of distinguishing subtle genetic aberrations of cells after they have been exposed to genotoxic materials and chemicals. This technique saves laborintensive work – a few hours versus the few weeks required by conventional methods. Although there is still much to be discovered about the screening method, it is clear that spectroscopy studies with FTIR and Raman would have potential as the initial evaluation methods to discover the characteristic genotoxic biomarkers to single out many materials, especially alloys for various BioMEM devices.

2.7.2 Surface Modification Protocols for Improved Biocompatibility and Applications in Biotechnology

As noted previously, the surface modification is an effective way to improve the biocompatibility of materials. Although there has been much progress made in the development of surface modifications technology, improving the biocompatibility of blood-contacting surfaces continues to be a challenge. For blood-contacting materials, perhaps the biggest challenge is to develop reliable surface modification methods to improve the blood compatibility of vascular grafts with a small diameter. Existing biomaterials have high risk of thrombogenic occlusion, which may causes the failure of cardiovascular devices. There are many surface modification approaches to creating nonthrombogenic surfaces. One approach involved coupling heparin to the material surface through physical, ionic and covalent binding (Barbucci et al., 2000; Vasudev et al., 2000). Another approaches involves biological modification of a material by albumin adsorption and seeding with endothelial cells (Bos et al., 1999). This approach is to immobilize biomembrane-mimetic surfaces to mimic the membranes of blood cells and avoid the recognition by blood as a foreigner. All these approaches showed improvement of blood compatibility of modified materials for different degrees. Nevertheless, there are currently no techniques available that can produce long-term nonthrombogenic surfaces. Attachment of endothelial cells would have been a good method if the cells could be tightly immobilized on biomaterials. Surface modifications with saline chemistry and thiolates are promising for creating nonthrombogenic surfaces *in vitro*, and further long-term *in vivo* study of surface - engineered materials is necessary to validate the efficacy of the surface modification protocols. Another promising approach to producing nonthrombogenic surfaces is to covalently immobilize adhesive proteins or peptides onto materials and promote the subsequent, adhesion of endothelial cells on the material.

A challenging in developing biosensors is to immobilize proteins on material surfaces while retaining their full functionality and stability. Most commonly used processes in immobilizing proteins on solid surfaces, for example, nonspecific adsorption, physical entrapment in polymeric gels, chemical cross-linking, or covalent attachment to an insoluble support, are difficult to control and usually yield randomly bounded proteins (Madoz et al., 1997). These bounded proteins may loss their natural conformation, and their activity is significantly reduced. One of the solutions to this problem is to develop new surface immobilization approaches that can use chemical functional groups with superaffinity to immobilize oriented proteins/ligands/DNA and remain their activity. For example, Spinke et al. (1993) introduced SAMs of biotinylated alkanethiols for binding streptavidin molecules. This biotin - avidin system can be subsequently used as a template to direct the binding of monobiotinylated Fab fragments of monoclonal antibodies. Recently, Kroger et al. (1999) used self-assembled monolayers of a synthetic chelator thioalkane which is capable of immobilizing oriented histidine-tagged proteins. They have demonstrated that the secondary structure of the protein was unaffected by the immobilization process. Further investigation in this field is to find out whether biotin – avidin can be applied to more universal proteins and to discover new affinity groups for various proteins to meet the growing needs in biotechnology.

References

Adamson, A. W., A. P. Gast. *Physical Chemistry of Surfaces*. John Wiley & Sons, Inc. (1997)

Amiji, M., K. Park. J. Biomat. Sci. Polym. Ed. 4:217 (1993)

An, Y. H., R. J. Friedman. J. Biomed. Mater. Res. 43:338 (1998)

Anderson, J. M. Soft tissue response. In: J. Black, and G. Hastings, eds., *Handbook of Biomaterial Properties*. Chapman & Hall, p. 490 (1998)

Andrade, J. D., V. Hlady. Protein adsorption and materials biocompatibility. Ins J. D. Andrade, L. I. Valuev, N. A. Plate, P. Klosinski, S. Penczek, and G. Glockner, eds., *Advances in Polymer Science*. Springer-Verlag Berlin Heidelberg, p. 3 (1986)

Bailey, F. E. Jr., J. V. Koleske. *Poly(ethylene oxide)*. Academic Press (1976)

Balachander, N., C. N. Sukenik, Amer. Chemical. Soc. 1 Langmuir. 11(6):1621 (1990)

Bambot, S. B., J. R. Lakowicz, G. Rao. Tibtech. 13:98 (1995)

Barbucci, R., A. Magnani, R. Rappuoli, S. Lamponi, M. Consumi. J. Inorg. Biochemi. 79 (1 – 4):119 (2000)

Bellantone, M., N. J. Coleman, L. L. Hench. J. Biomed. Mater. Res. 51(3): 484 (2000)

Black, J. *Biological Performance of Materials*. Marcel Dekker (1981)

Blawas, W. M. Reichert. Biomaterials. 19:595 (1998)

Bobyn, J. D., G. J. Wilson, D. C. Macgregor, R. M. Pilliar, G. C. Weatherly. J. Biomed. Mater. Res. 16:571 (1982)

Boone, C. W., N. Takeichi, S. A. Eaton, M. Paranipe. Science. 204:177 (1979)

Bos, G. W., N. M. Scharenborg, A. A. Poot, G. H. Engbers, T. Beugeling, W. G., Van Aken, J. Feijen. J. Biomed. Mater. Res. 44(3):330 (1999)

Brash, J. L. Thrombosis, hemostasis, and thrombolysis at prosthetic interfaces. In: M. Szycher. ed., *Biocompatible Polymers, Metals, and Composites*. Society of Plastics Engineers. 35 (1983)

Brash, J. L., Q. M. Samak. Dynamics of interactions between human albumin and polyethylene. J. Colloid Interface Sci. 65:495 (1978)

Brash, J. L. The physics and chemistry of protein-surface interactions. In: E. W. Salzman, ed. *Interaction of the Blood with Natural and Artificial Surfaces*. Dekker, 37 – 61 (1981)

Brody, J. P., T. D. Osborn, F. K. Forster, P. A. Yanger. Planar microfabricated fluid filter. Transducers. 1:779 (1995)
Bucher, V., M. Graf, M. Stelzle, W. Nisch. Biosensonsors & Bioelectronics. 14: 639 (1999)
Burns, N. L., J. M. Vanalstine, J. M. Harris. Langmuir. 11(7):2768(1995)
Chan, B., J. L. Brash. Interactions of human fibrinogen with glass surface. Thromb. Haemostas. 42:132 (1979)
Chehroudi, B., T. R. L. Gould, D. M. Brunette. J. Biomed. Mater. Res. 26:493 (1992)
Chen, C. S., M. Mrksich, S. Huang, G. M. Wihitesides, D. E. Ingber. Biotechnol. Prog. 14:356 (1998)
Chirila, T. V., D. E. Thompson, I. J. Constable. J. Biomater. Sci. Polym. Ed. 3: 481 (1992)
Chou, L., J. D. Firth, V. J. Uitto, D. M., Brunette. J. Cell Sci. 108:1563 (1995)
Chou, L , J. D. Firth, V. J. Uitto, D. M., Brunette. J. Biomed. Mater. Res. 39 (3):437 (1998)
Christ, F. R., S. Y. Buchen, et al. Biomaterials used for intraocular lenses. In: *Encyclopedic Handbook of Biomaterials and Bioengineering*. Part B. ed. by D. L. Wise. and D. J. Trantolo et al. Marcel Dekker, New York, p. 1361 (1995)
Collings, A. F., F. Caruso. Rep. Prog. Phys. 60:1397 (1997)
Creighton, T. E. *Proteins*. W. H. Freeman (1993)
Curtis, A., C. Wilkinson. Biomaterials. 18:1583 (1997)
Curtis, A., C. Wilkinson. J. Biomat. Sci. Polym. Ed. 9(12):1313 (1998)
Desai, T. A., W. H. Chu, P. Sinibaldi-Vallebona, P. Borboni, G. Beattie, A. Hayek, M. Ferrari. Implantation of microfabricated immunoisolating biocapsules, Micro and nanofabricated electro-optical-mechanical systems for biomedical and environmental application. In: P. Gourley, ed. *SPIE Proceeding Series*, 40 (1998)
Elwing, H. S., A. Welin, U. Askendal, I. A. Nilsson, J. Lundstrom, Colloid Interface Sci. 119(1):203 (1987)
Feldman, D. S., S. M. Hultman, R. S. Calaizzo, A. F. Von Recum. Biomaterials. 14:105 (1983)
Feng, L., J. D. Andrade. Biomaterials. 15(5):323 (1994)
Ferrari, M., W. H. Chu, T. A. Desai, D. Hansford, M. Zhang. MRS. 414:101 (1996)
Flounders, A. W., D. L. Brandon, A. H. Bates. Appl. Biochem. Biotechnal. 50:265 (1995)
Green, R. J., R. A. Frazier, K. M. Shakesheff, M. C. Davies, C. J. Roberts, S. J. B. Tendler. Biomateials. 21:1823 (2000)
Gristina, A. G., M. Oga, L. X. Webb, C. Hobgood. Science. 228:990 (1985)
Groves, J. T., N. Ulman, S. G. Boxer. Micropatterning fluid lipid bilayers on solid supports. Science. 275:651 (1997)
Harris, A. Exp. Cell Res. 77:285 (1973)
Harris, J. M. *Biotechnical and Biomedical Applications*. Plenum Press, p. 303 (1992)

Helmus, M. N., K. Tweden. Materials Selection. In: D. L. Wise, D. J. Trantolo, D. E. Altobelli, M. J. Yaszemskim, J. D. Gresser, and E. R. Schwartz, ed. *Encyclopedic Handbook of Biomaterials and Bioengineering*. Marcel Dekker, New York 1:27 (1995)

Hengerer, A., J. Decker, E. Prohaska, S. Hauk, C. Koblinger, H. Wolf. Biosensors & Bioelectronics. 14:139 (2000)

Holman, H. N., M. Zhang, R. G. Goldstein, M. Ferrari, J. C. Hunter-Cevera. SPIE, BIOS'. 3606:55 (1999)

Holmberg, K., C. Bergstrom, M. B. Stark. Biotechnical and Biomedical Applications, Pleum Press, 303 – 324(1992)

Horbett, T. A., J. L. Brash. *Proteins at Interfaces: Physicochemical and Biochemical Studies*. American Chemical Society, Washington, DC., p. 343 (1987)

Horbett, T. A., B. D. Ratner, J. M. Schakenraad, F. J. Schoen. *Some Background Concepts Biomaterials Science*. ed. by B. D. Ratner, and A. S. Hoffman. Academic Press, p. 133 (1996)

Hsiue, G. H., S. D. Lee, P. C. Chang. J. Biomat. Sci. Polym. Ed. 7(10), 839 (1996)

Huang, J., D. A. Dahlgren, J. C. Hemminger. Langmuir. 10:626 (1994)

Hunt J. A., G. Meijs, D. F. Williams. J. Biomed. Mater. Res. 36:542 (1997)

Ito, Y. Biomaterials. 20:2333(1999)

Jenny, C. R., J. M. Anderson. J. Biomed. Mater. Res. 44:206 (1999)

Johnson, H. J., S. J. Northrup, P. A. Seagraves, M. Atallah, P. J. Garvin, L. Lin, and T. D. Darby. J. Biomed. Mater. Res. 19:489(1985)

Kajiwara, Y., S. Ajimi, A. Hosokawa, K. Maekawa. Mutation Res. 293: 81 (1997)

Kilbey, B. J., M. Legator, W. Nichols, C. Ramel. *Handbook of Mutagenicity Test Procedures*. Elsevier Scientific (1977)

Kittilsland, G., and G. Stemme. Sensors and Actuators. 1:904 (1990)

Kooten, T G. V., J. F. Whitesides, A. F. V. Recum. J. Biomed. Mater. Res. 43:1 (1998)

Kroger, D., M. Liley, W. Schiweck, A. Skerra, H. Vogel. Biosensors & Bioelectronics. 14:155(1999)

Lahiri, J., E. Ostuni, and G. M. Whitesides. Langmuir. 15:2055 (1999)

Lanza, R. P., W. L. Chick. Sci. Am. 16 (1995)

Lee, W. E., H. G. Thompson, J. G. Hall, D. E. Bader. Biosensonsors & Bioelectronics. 14:795 (2000)

Li, J. T., J. Carlsson, J. N. Lin, K. D. Caldwell. Bioconjugate Chem. 7:592 (1996)

Liu, C., Z. Jin. Tibtech. 15:213 (1997)

Lom B., K. E. Healy, P. E. Hockberger. J. Neurosci. Methods. 50:385 (1993)

Madoz, J., B. A. Kuznetzov, F. J. Medrano, J. L. Garcia, V. M., Fernandez. J. Am. Chem. Soc. 110:1043 (1997)

Malmsten, M., B. Lassen. Ellipsometry studies of protein adsorption at hydrophobic surfaces. In: Horbett, T. A. and J. L. Brash, eds. *Proteins at Interfaces II*.

American Chemical Society, Washington, D. C., p. 288 (1995)
Mandenius, C. F., S. Welin, B. Danielsson, I. Lundstrom, K. Mosbach. Anal. Biochem. 106 (1984)
Marrazza, G., I. Chianella, M. Mascini. Disposable DNA electrochemical sensor for hybridization detection. Biosensonsors & Bioelectronics. 14:43 – 51 (1999)
Martin, P. M., D. W. Matson, W. D. Bennett, D. C. Stewart, Y. Lin. SPIE. 3680:826 (1999)
McPherson, T. B., S. J. Lee, K. Park. Analysis of the prevention of protein adsorption by steric repulsion theory. In: T. A. Horbett and J. L. Brash, eds. *Proteins at Interfaces* Ⅱ. American Chemical Society, Washington, D. C., p. 395 (1995)
Metcalf, B. W., B. J. Dalton. *Cellular Adhesion.* Plenum Press (1994)
Meyele, J., A. F. von Recom, B. Gibbesch, W. Huttenmann, U. Schlagenhauf, W. Huttenmann, U. Schlagenhauf, W. W. Shulte. J. Appl. Biomater. 2:273 (1991)
Morgenthaler, J. J. *Virus Inactivation in Plasma Products.* Karger, New York (1989)
Morgenthaler, J. J. Inactivation and partitioning of viruses in an integrated plasma fractionation process. FDA/NHLBI Workshop, 1 – 2 (1994)
Morpurgo, M., F. M. Veronese, D. Kachensky, J. M. Harris. Bioconjugate Chem. 7:363 (1996)
Mrksich, M. , Whitesides. Annu. Rev. Biophys. Biomol. Struct. 25:55(1996)
Mrksich, M., L. E. Dike, J. Tien, D. E. Ingber, G. M. Whitesides. Exp. Cell Res. 235:305 (1997)
Mulchandani, P., A. Mulchandani, I. Kaneva, W. Chen. Biosensor for direct determination of organophosphate nerve agents. 1. Potentiometric enzyme electrode, Biosensors & Bioelectronics. 14:77 – 85 (1999)
Nagase, M.. Host reaction to particulate biomateials. In: D. L. Wise, D. J. Trantolo, D. E. Altobelli, M. J. Yaszemski, J. D. Gresser, and E. R. Schwartz, eds. *Encyclopedic Handbook of Biomaterials and Bioengineering.* Marcel Dekker, Vol. 1. New York 269 (1995)
Neville, A. C.. *Biology of Fibrous Composites.* In: A. C. Neville, ed. Cambridge University Press (1993)
Nygren, H., M. Stenberg. J. Biomed. Mater. Res. 22:1 (1988)
Pale-Grosdemange, C., E. S. Simon, K. L. Prime, G. M. Whitesides. J. Am. Chem. Soc. 113:12 (1991)
Park, K. D., Y. S. Kim, D. K. Han, Y. H. Kim, E. H. B. Lee, H. Suh, K. S. Choi. Biomaterials. 19:851 (1998)
Peluso, G., O. Petillo, J. M. Anderson. J. Biomed. Mater. Res. 34:327 (1997)
Peppas, N. A., R. Langer. Science. 263:1715 (1994)
Peters, T. Jr. ed. *All about Albumin.* Academic Press, New York (1996)
Pilliar, R. M.. Clin. Orthop. 176:42 (1983)
Pizzoferrato, A., C. R. Arciola, E. Cenni, G. Ciapetti, D. Granchi, L. Savarino, S. Stea. D. L. Wise, D. J. Trantolo. In: D. L. Wise, D. J. Trantola, D. E. Altobelli, M. J. Yaszemski, J. D. Gresser, and E. R. Schwartz eds.

Encyclopedic Handbook of Biomaterials and Bioengineering. Marcel Dekker, 1:329 (1995)

Pizzoferrato, A., C. R. Arciola, E. Cenni. Biocomptibility of implant materials: a quantitative approach. In: D. L. Wise, ed. *Encyclopedic Handbook of Biomaterials and Bioengineering*. Marcel Dekker, pp. 329 – 370 (1995)

Pratten, M. K., J. B. Lloyd. Biochem. Biophys. Acta. 881:307 (1986)

Prime, K. L., Whitesides. G. M. J. Am. Chem. Soc. 115:10714 (1993)

Putnam, F.. The Plasma Proteins: Structure, Function and Genetic Control. New York, Academic Press (1975)

Raay, J. J. A. M. V., P. M. Rozing, C. A. V. Blitterswijk, R. M. V. Haastert, H. K. Koerten. J. Mater. Sci.: Mater. Medi. 6:80 (1995)

Ramsden, J. J. Biosensors & Bioelectronics. 11(5):523 (1996)

Ratner. B., A. S. Hoffman, F. J. Schoen, J. E. Lemons. Biomaterials Science. Academic Press, New York (1996)

Reid, J. Mater Sci. 201 (1993)

Samsonova, J., M. Y. Rubtsova, A. V. Kiseleva, A. A. Ezhov, A. M. Egorov. Biosensors & Bioelectronics. 14:273 (1999)

Schakenraad, J. M. Cells: their surfaces and interactions with materials. In: B. Ratner, A. S. Hoffman, F. J. Schoen, and J. E. Lemons, eds. *Biomaterials Science: Introduction to Materials in Medicine*. Academic Press, p. 141 (1996)

Schmalzried, T. P., W. J. Maloney, M. Jasty, L. M. Kwong, W. H. Harris. J. Arthroplasty. 8:179 (1993)

Schugerl, K., B. Hitzmann, H. Jurgens, T. Kulick, R. Ulber, B. Weigal. Tibtech. 14 (1996)

Shanbhag, A. S., J. J. Jacob, J. Black, J. O. Galante, T. T. Glant. J. Biomed. Mater. Res. 28:81 (1994)

Silver, F. H.. *Biomaterials: Medical Devices and Tissue Engineering*. Chapman & Hall (1994)

Silver, F. H., D. L. Christiansen. *Biomaterials Science and Biocompatibility*. Springer-Verlag, New York (1999)

Silver, J. H., R. W. Hergenrother, J. C. Lin, F. Lim, H. B. Lin, T. Okada, M. K. Chaudhury, S. L. Cooper. J. Biomed. Mater. Res. 29:535 (1995)

Singh, J., A. R. Ray, J. P. Singhal, H. Singh. Biomaterials. 473 (1990)

Smith, B., M. V. Sefton. J. Biomed. Mater. Res. 27(1):89(1993)

Spinke, J., M. Liley, F. J. Schmitt. J. Chem. Phys. 99(1):7012(1993)

Stenberg, M., H. Nygren. J. of Physique. C10:83(1983)

Syrjanen, S., A. Hensten-Pettersen, K. Nilner. J. Biomed. Mater. Res. 20:1125 (1986)

Takaoka, T., M. Okumura, H. Ohgushi, K. Inoue, Y. Takakura, S. Tamai. Biomaterials. 17:1499 (1996)

Tani, N. Organs. 20(8):922 (1996)

Tender, L. M., R. I. Worley, H. Fan, G. P. Lopez. Langmuir. 12:5515(1996)

Tolles, W. M. Nanotechnology. 7:59 (1996)

Tompkins, H. G. A user's guide to ellipsometry. Academic Press (1993)
Tredgold, R. H.. Order in thin organic films. Cambridge University Press (1994)
Tu, J., T. Huen, R. Szema, M. Ferrari. In Gourley, P. Ed. *Micro and Nanofabricated Electro-Optical Mechanical Systems for Biomedical and Environmental Application.* SPIE Proceeding Series. 148 – 155(1998)
Ulman, A. *An Introduction to Ultrathin Organic Films.* Academic Press, New York (1991)
Van-Kooten, T. G. V., J. F. Whitesides, and A. F. V. Recum. J. Biomed. Mater. Res. 43:1(1998)
Van-Kooten, T. G. V., J. F. Whitesides, and A. F. V. Recum. J. Biomed. Mater. Res. 43:1(1998)
Van-Kroger, D., M. Liley, W. schiweck, A. Skerra, H. Vogel. Biosensors and Bioelectronics. 14:155(1999)
Vasudev, S. C., L. R. Moses, C. P. Sharma. Aritifi. Cells Blood Substitutes Immobilization Biotechnol. 28(3):241 (2000)
Veiseh, M., K. Hinkley, Y. Zhang, M. Zhang. Two-dimensional protein micropatterning for sensor applications through chemical selectivity technique. Biomedical Microdevices. In press (2000)
Verkleij, M. W., L. F. Morton, C. G. Knight, P. G. De Groot, M. J. Barnes, J. J, Sixma. Blood. 91(10):3808 (1998)
Voet, D.. *Biochemistry.* John Wiley & Sons (1995)
von Recum, A. F., J. B. Park. Bioengineering. 5:37 (1981)
Walivaara, B., A. Askendal, H. Elwing, I. Lundstrom, P. Tengvall. J. Biomed. Mater. Res. 26:1205 (1992)
Walivaara, B., P. Warkentin, I. Lundstrom, P. Tengvall. J. Colloid and Interface Sci. 174:53 (1995)
Wallin, R. F., P. J. Upman. Evaluating the biological effects of medical devices and materials. In: D. L. Wise, ed. *Encyclopedic Handbook of Biomaterials and Bioengineering.* Marcel Dekker, p. 399 (1995)
Wannerberger, K., T. Arnebrant. J. Colloid and Interface Sci. 177:316 (1996)
Wening, J. V., H. Marquardt, A. Katzer, K. H. Jungbluth. Biomaterials. 16(4):337 (1995)
Whitesides, G. M. *Scientific American.* 146 – 149(1995)
Williams, R. A., H. W. Blanch. Biosensors and Bioelectronics. 9:159 – 167 (1994)
Williams, D., R. L. Williams. Degradative effects of the biological environment on metals and ceramics. In: B. D. Ratner, A. S. Hoffman, F. J. Schoen, and J. E. Lemons, eds. *Biomaterials Science: Introduction to Materials in Medicine.* American Press, p. 260 (1995)
Wise, D. L., D. J. Trantolo, D. E. Altobelli, M. J. Yaszemski, J. D. Gresser, E. R. Schwartz. *Encyclopedic Handbook of Biomaterials and Bioengineering*, Vol. 1. Marcel Dekker (1995)
Wood, E. M., E. Colton, R. A. Yomtovian, L. M. Currie, J. Connor, J. M. Anderson. Outstanding Research by a Resident or Clinical Fellow. World Biomaterials Congress 2000. ed. by H. I. Kamuela. John Wiley & Sons,

p. 147 (2000)
Woodhouse, K. A., J. I. Weitz, J. L. Brash. JBMR. 28(4):407 (1994)
Woolley, A. T., R. A. Mathies. Proc. Natl. Acad. Sci. USA 91:11348 (1994)
Wyatt, D., J. Keathley, C. Williams, Broce. Inactivation of biological contaminants. BioPharm. 34 – 40 (1993)
Yoda. R Biomater. Sci: Polymer Edn. 9(6): 561 – 626 (1998)
Zalipsky, S.. Bioconjugate Chem. 6:150 (1995)
Zhang, M., M. Ferrari. 1(1):81 (1998a)
Zhang, M., M. Ferrari. Biotech. & Bioeng. 56(6):618 (1998b)
Zhang, M., M. Ferrari. SPIE. 3258:15 (1998c)
Zhang, M., H. N. Holman, M. Ferrari, J. C. Hunter-Cevera. *Mat. Res. Symp.* Proc. 550:163 (1999)
Zondervan, G. J., H. J. Hoppen, A. J. Pennings, W. Fritschy, G. Wolters. Biomaterials. 13(3):137 (1992)

3 Biotechnological Applications of Inorganic Glasses

Alexis Clare

3.1 The Nature of Inorganic Glass

Rarely thought of as a biomedical material, inorganic glass is ubiquitous in biotechnology and has the potential of being an extremely useful and very flexible biomedical material. When one thinks of glass, it is usually in terms of its application, for example, window glass and drinking glass. However glass is in fact a whole subgroup of materials under the classification "solid state".

Just as matter is divided into solids, liquids, and gasses, glases are materials that behave like an elastic solid but have an atomic structural arrangement that is similar to the liquids from which they are formed (apart from a few exceptions such as sol-gel materials, see later). While adhering to the same bonding constraints and, therefore, local (0 to 5 Å) order as their crystalline counterparts, the structure has no long-range (> 10 Å), and limited medium (5 – 10 Å) range order, and certainly no periodicity. It is this lack of periodicity and long-range order that makes many inorganic glasses transparent in the visible region of the spectrum.

In network glasses, chemical compounds are grouped according to the way in which they affect the glass structure. A glass "network former" is a compound that, if cooled at a reasonably rapid rate, will, on its own, form a glass. A glass "network modifier" is a compound that will not form a glass on its own and when added to a network former compound will break up the network bonds creating nonbridging anions. An "intermediate" is a compound that, on its own will not form a glass but depending upon the other components may take on a former or modifier role in the structure.

A vast majority of network glasses and the inorganic glasses with which one is most familiar, are oxide glasses. A majority (> 90%) of all commercial glasses are oxide and greater than 90% of those are based upon silica. Silica is the archetype network former with soda and lime being network modifiers. Silica is the principal component in optical fiber for telecommunications and some sensing applications with only trace amounts of secondary components. A large proportion of bulk commercial glass is based upon a soda – lime – silica composition. while others may be alumino or borosilicate such as some fiberglass and labware. Commercial silicate glasses feature strongly in the biotechnology industry, as will be discussed later, however, one range of soda – lime – silicate compositions has made its mark as a biocompatible material for devices: Bioglass® invented by Larry Hench.

Glasses do not have to contain silica. One can make glass from phosphates, borates, and even germanates, vanadates, and gallates! They can even be made to contain large quantities of rare earth as in the case of Delbert Day's therapeutic glass spheres. Glasses do not have to contain even oxygen. Ionic glasses consisting of only fluorides or chlorides can be made. These glasses often do not form traditional networks like the oxides (with the exception of beryllium fluoride) but form high coordination number glasses with many components. These glasses have found niches in biotechnology for sensing [predominantly in the infrared (IR)], in medicine for the delivery of high power midinfrared laser light, and, in some cases, high-power UV laser light. Glasses containing arsenic, germanium, and antimony coupled with sulfur, selenium, and tellurium are polymeric inorganic structures called chalcogenide glasses. These glasses are structurally interesting but also have the capacity to transmit long wavelength IR light for laser treatment whether as an optical fiber conduit or as a lens. Finally, even metals, when cooled rapidly enough, can form glasses. There are occasions where the absence of grain structure in a material might prove advantageous in the field of biotechnology and medicine.

3.2 Why Inorganic Glasses?

There are many advantages to glass for the field of biotechnology. Glass is an extremely flexible and tailorable material to use in this field.

3.2.1 Homogeneity

Unlike polycrystalline materials, most glasses are homogeneous. Polycrystalline materials consist of grains which may have compositional variations as well as variations in crystal phase. In between the grains are the grain boundaries that can have an amorphous phase, impurities, and defects present. The microsctructure of the material carries through to the surface and can affect the way in which an organism will interact with that surface. Glass has no grains or grain boundaries, it can, however, be inhomogeneous on occasion. For example, some glass compositions undergo liquid – liquid phase separation which is a phenomenon that can have a variety of morphologies associated, for example, droplet formation or interconnecting phases. A combination of the latter with leaching leads to the Vycor® process in which a borosilicate glass is phase separated and the leachable phase is dissolved, leaving a porous glass that is 97% silica.

3.2.2 Compositional Flexibility

Of the 112 elements in the periodic table, there are only about 20 that have not been incorporated yet into a glass, and this is mainly because the elements are so rare that they are not often incorporated into anything. The quantity and sheer number of things that can be incorporated into glass is one of its true advantages; it is why it is a favorable nuclear waste medium. When processing crystalline materials, an accident of stoichiometry can transport one into a region of a phase diagram for which an undesirable microstructural evolution might occur. Glass, on the other hand, is not inhibited by stoichiometry, and provided that a compositional variation does not extend one into a region of devitrification or phase separation, the addition of a given component often leads to a linear variation in the properties of the glass. The latter is not always true. For example, if one mixes alkali ions in the glass, one sees a curious phenomena known as the "mixed alkali effect" in which a minimum or maximum in a given property over a specified range might be observed. Nevertheless, the ability of glasses not only to be continuously varied over these ranges, but also to accommodate a wide variety of components, makes them much more tailorable in terms of their properties.

3.2.3 Flexibility in Manufacturing

There are a number of ways to make glass. The most traditional is the melt quench technique in which raw materials, which are usually the crystalline components of the glass, are mixed intimately and taken to high temperature (not always necessarily above the melting point of the highest melting component because lower melting components will often act as a flux for the higher melting components) and then cooled quickly to a temperature below the glass transition. If one reheats a glass, it does not "melt" in the same way that a crystalline materials does. It undergoes a transition from behaving like an elastic solid to behaving like a viscous liquid; that is known as the glass transition region. Unlike a melting point, the glass transition is not constant, but is dependent on the thermal history of the glass. The speed at which cooling must proceed depends very much on the type of glass; one can make a glass from molten silica which can undergo very slow cooling without crystallization, whereas making a glass from a metal requires cooling rates in excess of millions of degrees per second.

One of the important features of glass melting and formation is the flexibility in the geometry of the final piece. It is possible to cast glass into almost any type of monolithic shape including glass fibers, microspheres, or powders. Glass fibers have many applications including strengthening of composites (thin fibers, a few microns in diameter) to forming fiber optic sensors and endoscopes (thicker fibers of a hundred or so microns in diameter with a high refractive index core/low refractive index cladding to guide light). Glass microspheres have been proposed for all types of uses from sensing, to cancer treatment, to purification of DNA. Powders can be used in wound healing, dental applications, and purification.

There are other methods of glass-making, too, that do not involve high temperatures, for example, the sol-gel method in which organic precursors are reacted together usually by hydrolysis and condensation to form a glass. The formation of a solid monolith by this method is quite difficult because the drying process, unless carried out very carefully, can lead to cracking. On the other hand, coatings are made regularly by this method. The most interesting aspect of this method from a biotechnology standpoint is the possibility of encapsulating, in the sol-gel structure, biomolecules such as proteins, which may be used for biomedical purposes or sensing purposes. In addition, the sol-gel method tends to lead to porous samples which under certain conditions can be controlled to allow molecular filtering of cell ingrowth. Porosity can be introduced by other methods also such as by sintering glass particles with a filler that will be removed on heating or by dissolution. Alternatively, porous glass can be made by the plasma spray technique where glass particles are coated onto a substrate and porosity is introduced by spraying under different conditions. However, it is often the case that the desirable bulk properties of a material are not the same as the required surface interactions, and surface modification is something to which glasses are quite amenable.

3.2.4 Surface Modification

It has already been stated that the glass transition of a composition is dependent upon the thermal history, this also extends to the structure of the elastic solid; particularly the volume. Cooling glasses at different rates freezes in different structures. As a result, if the cooling rates between the interior and exterior differ considerably, then the structures will also differ. This is demonstrated quite neatly by the response of the side window of an automobile to a sharp implement. The window shatters catastrophically into small, almost cubed pieces but only under sharp indentation with some force. This glass is tempered for strength and safety. Glass is not the first material that one thinks of as "strong", but the theoretical strength of many glasses is very high; the weakness comes from flaws, mostly surface flaws. Hence, treatments that heal or close flaws on the surface tend to make glass stronger. One such treatment is ion exchange. If a glass contains mobile ions, for example, sodium in a soda-lime-silica (e.g. window glass, microscope slide) glass, those ions can be leached out and replaced by ions that are chemically similar but much larger in size. Since this is usually carried out at temperatures below the glass transition temperature, the surface that now contains the larger ions is under compression. This closes the flaws and prevents them from propagating. This technique has been used for strengthening glass in the biological arena (e. g. glass vials that contain the medicine used in emergency injection devices for persons with severe allergies), but in addition to strengthening, it can chemically alter the glass in terms of the surface response.

There are a number of techniques such as coating that can change the functionality of glass surfaces. LaCourse (1996) and co-workers showed that ammonia treatments of a bioresorbable phosphate glass fiber could cause two orders of magnitude change in the dissolution rate of the fiber. Common sterilization procedures for materials can change the surfaces quite significantly: autoclave treatments leach mobile ions out of a glass structure leaving them with a very durable network glass layer. Gamma sterilization can cause defects in the glass, which can make the surfaces quite reactive. Glow discharge treatments have even been shown to enhance bone integration in bone-bonding glass-ceramic implants (Muller-Mai et al., 1992). In short, there is as much flexibility in the surface engineering of a glass as there is in the bulk engineering to get desirable properties.

3.2.5 Other Modifications

As stated previously, glasses are peculiarly dependent structurally on their thermal history. If one processes a glass by cooling from the melt, one can introduce uneven stress patterns in the glass piece mainly because glasses have rather poor thermal conductivity. Hence, if different sections cool at different rates, then different structures will be frozen in, and the mismatches in those structures result in strain. This is easily seen if one looks at a glass made in this way by placing it between crossed polars. Fringes are seen particularly in the regions of high strain. This strain can lead to a sudden and catastrophic explosion of the glass which is beautifully demonstrated if you have ever seen the fracture of a tempered glass window. Annealing is a process whereby slow reheating to just above the glass transition and slow cooling of the glass can alleviate the mismatched stresses. However, annealing also changes the glass structure. Glass fibers are a fast-cooled glass that does not usually have to be annealed because the size means that heat is removed fast and fairly evenly. Nevertheless, the cooling is fast, and therefore, the structure of the fiber is considerably different from that of a bulk glass of ostensibly the same composition. Choueka et al. (1995) examined some bioresorbable glass fiber designed for bone fixation composites (see later) and saw that annealing significantly changes the chemical durability. This effect is really only observed in a glass and is another testament to the flexibility in structure that can be achieved just by relatively simple treatments.

3.3 Properties Pertinent to Biotechnology and Biomedicine

If one considers the whole world of biotechnology and medicine, there is probably an application for every property in existence; nevertheless, in this section, the properties more pertinent to biotechnological and medical applications will be discussed. Probably the most important of these is chemical durability.

3.3.1 Chemical Durability

There are requirements in biotechnology and biomedicine for a whole range of chemical durability behavior. There exists quite a large database of behaviors of different glass compositions in various media. If one starts with silicate glasses, those with high network former content tend to be more chemically durable than those with a high modifier content. For example, the Bioglass® compositions tend to have a silica content that is lower than 60 mol% and as a result, undergo a reasonably fast mechanism in which leaching of mobile ions is followed by formation of a silica gel layer, onto which precipitates of calcium and phosphorus can form an amorphous calcium phosphate film. This film will crystallize into a hydroxyapatite carbonate that is not dissimilar to that found in bone. The result of this is that compositions of Bioglass® have better bone bonding properties than most laboratory-made hydroxyapatite materials. In fact Kokubo (2000) argues that for any material given a sufficient proportion of OH groups at the surface, a bond will form with bone in the body environment where there is an abundance of calcium and phosphorus to form the amorphous calcium phosphate that is the precursor to hydroxyapatite. The advantage of Bioglass-like compositions is that the process is faster because of the surface reactivity and, Bioglass® itself contains calcium and phosphorus, which can add to the solution and assist with the precipitation. In addition, Bioglasses®, unlike other bioactive materials, form bonds with soft tissue as well as hard tissue. Bioactive glasses have found many applications in wound healing both in bone and soft tissue, middle ear implants, and in dentistry. There are some good reviews of bioactive glasses, for example, Hench and West (1996) and Hench and Wilson (1999). Most of the applications for these materials have been nonload-bearing. As will be shown later, one way to make glass a load-bearing material is to turn it into a glass-ceramic. In a glass, ceramic, strength is increased by the presence of small crystallites in the structure, which block crack growth. Some examples of these are the apatite – wollastonite glass-ceramics that have been successfully used in spinal defect repair in Japan (see later). Glass ceramics can allow one the flexibility of glasses in terms of compositional variations but with added mechanical properties as a result of the devitrification step. Sometimes the devitrification step may be used to introduce another property into the glass such as magnetic properties (see later).

There are some glasses in the Bioglass® ternary and other types of glasses such as some phosphate compositions and some borate compositions that disappear altogether after a designated period of time. These are known as bioresorbable. There are many reasons why one might wish to design a bioresorbable glass; it has been mentioned previously that glass is strong, provided surface flaws are inactivated. Glass fiber is extremely strong when first formed and can maintain its strength if it is coated immediately after formation. Hence, glass fiber composites can be formed. Recreational sailors and even some motorists are fully apprised of the strength to weight ratios of glass fiber composites, and there are many conceptual designs for glass fiber reinforced bioresorbable composites for bone repair. The advantage of such a system is the alleviation of some of the problems resulting from bone remodeling (Lin et al., 1994; Marcolongo et al., 1996). If bone injuries are fixed with a nonresorbable device, the bone may atrophy around the device if the device assumes the load for too long. A bioresorbable device would not require surgical removal and, if designed appropriately, could gradually let the healing bone assume the load in a more natural fashion. Other reasons why bioresorbability might be desirable would be drug delivery or therapeutic treatments. The key to a successful bioresorbable product is to ensure that the products of the dissolution do not cause problematic imbalances in the biological system, for example, sudden pH changes that might overcome a local buffer or a systemic overload of a system by an electrolyte. This again is one of the beauties of a glass system in that one can introduce a buffer in the structure that can help avoid these issues as well as control the dissolution rates to a high degree of accuracy.

At the opposing end of the chemical durability spectrum is the durable material and the fact that one still finds glass from ancient times, when metals and polymer (e. g. paper) have long gone, is a testament to the durability of glass. Glass can be made so durable that in many countries it is the material of choice for storage of nuclear waste. With a few exceptions, glasses with high durability are those that have to be fabricated at high temperatures and which have a resulting high glass transition temperature. In network glasses, this is usually brought about by having a large quantity of glass former present with only small amounts of modifier. In addition, to ensure a durable glass, it is usual to have more than one type of modifier ion. The presence of two modifiers acts to inhibit each other's diffusion, limiting the leaching of the glass.

The development of pharmaceutical glasses, which are designed to be completely inert to a wide variety of chemicals, is a rigorous and involved process. Particularly in the field of biotechnology, there has been a tendency to look at glass as a single material rather than a state of matter and while many of our food and beverage containers are soda lime silica glass, special precautions need to be followed to ensure that a glass is totally inert. Soda lime silica containers will undergo some leaching of mobile ions into solution. This occurs even when one does not think one has an aqueous solution. It has been demonstrated that there is enough water absorbed into alcohol to observe leaching of sodium ions (Scholze, 1990). For the average beer and wine bottle, this is not significant (and most glass manufacturers insert a small amount of alumina in their glass batches to limit the sodium dissolution), but it can be an issue particularly in pharmaceuticals where adverse reactions may occur, as defined in the European and American Pharmacopoeia. USP and FE type I glasses are the most inert and are generally a borosilicate glass with very high thermal shock resistance to withstand sterilization. These containers are resistant to acid, alkali, and neutral pH. USP and FE type II is a treated soda lime silicate glass which on the treated surface is resistant to acid and neutral pH, usually more so than type I. Usually, type II is dealkalized, for example, by an SO_3 treatment (Hubert, 1986) and hence is not base-resistant. These containers can be amber-colored for light-sensitive products. USP and FE type III are usually low alkali soda lime silicate glasses for nonaqueous and dry products. Lower thermal shock resistance usually limits these containers to dry heat sterilization. New developments in pharmaceutical glass include the use of ion exchange strengthening to increase the strength to weight ratio for special applications (e.g. the aforementioned emergency inoculations). Other high durability applications include sensors and diagnostics.

Glass is the material of choice if a sensor is based upon optics. The optical quality and mechanical stability of polymers for optical based sensors are just not up to the quality of glass. Hence, glass optical fiber is widely used in medical imaging for optical fiber endoscope bundles and in laser surgery as a delivery device. It is also used in both passive and active optical sensors. In these applications, it is essential that the glass be as inert as possible.

Another medical application of durable glass, which is now in clinical use in the United States, is the radioactive glass beads developed by Day at the University of Missouri Rolla (Day, 1995a, 1995b; McIntyre and Day, 1998a). These beads are made from rare earth aluminosilicate glasses that are very durable because shortly after they are manufactured, they are inserted into the core of a nuclear reactor to activate the rare earth isotopes. Examples include dysprosium and samarium isotopes, which, on activation, become betaemitters. Beta radiation has a very short path length in most materials and hence if these beads are implanted into a tumor, the radiation only affects the tumor locally. In addition, the half-lives of the isotopes which become active are quite short (hours – days) so that the radioactivity of the beads dies down within a reasonable length of time. Thus, one can treat cancer with a high dose of radiation locally. This method has been used for the treatment of liver cancer, which is difficult to treat by other means. The increased vascularity of the tumors in the liver cause the beads to flow preferentially to the tumor site, delivering the radiation dose precisely where it is required. In this case, one must ensure that the glass is durable over a period longer than the above-background activity of the isotope.

The presence of passive glass beads in the liver has had no known adverse effects on treated patients; however, one other application of similar beads requires a rather different chemical durability modality. It is known that radiation delivered to the synovium can help sufferers of rheumatoid arthritis. A localized treatment similar to that described for the liver is proposed for synovectomy (McIntyre and Day, 1998b). However, unlike the liver, continued presence of glass beads in the synovial fluid would cause severe aggravation. A different type of radioactive bead is proposed where the bead does not dissolve while the activity is still present but that the beads dissolve in a controlled manner once the activity has fallen below background levels. In this case a rather sophisticated design of the chemistry is required with functional layers in which the chemical durability differs.

3.3.2 Surface Chemistry

The simplest network glass is silica. Silica is a network of SiO_4 tetrahedra that are shared only at the corners through the oxygen which is, as designated previously, the bridging oxygen. The surface of silica glass, however, is a series of dangling bonds that, provided it is not a fresh surface newly created in a high vacuum, will be coordinated by OH groups from the moisture in the air. These Si – OH groups are called silanols. Once one starts to add other components, the surface will contain other metal – OH groups. However, the proportions of other components in the surface may not be the same as their proportions in the bulk. In a molten glass, before formation, the surface composition is controlled by the atmosphere surrounding the glass and the surface tension of the glass. The melt will adjust the composition to lower the surface tension. Hence, if your reactive surface is a result of the formation from the melt, then you will have possibly a quite different surface chemistry to that of the melt. This is likely in fibers, flat glass, beads, and monolithic shapes. The surface of powdered glass will depend on the method used to grind up the glass. Now, of course, it is possible to modify the surface chemistry, as previously alluded to, but it is important to analyze the surface first to know exactly what it is that one is analyzing.

The surface chemistry and surface interactions are obviously very important for bioactive and bioresorbable glasses, and they have been reasonably well studied using a variety of techniques such as Fourier transform infrared reflectometry (FTIR) and secondary ion mass spectrometry (SIMS). But newer techniques being developed such as near field scanning optical microscopy (NSOM) show promise of revealing much more about molecular interactions at these interfaces. The real importance of the interface will be in the realm of molecular attachment. The relatively modern field of tissue engineering is definitely encouraging better understanding of the mechanisms by which biomolecules attach to substrates, and I believe there will be a place for glass materials in tissue engineering in the future because of the flexibility mentioned before.

Molecular attachment is also very important in the biotechnological diagnostics and assay fields. There are a number of well-developed technologies for molecular diagnostics that are predominantly based upon two-dimensional arrays of oligonucleotide probes that can screen large numbers of assorted genes (Chee et al., 1996). Applications for this technology range from human genomics to home diagnostics. The methods vary somewhat; for example, Affymetrix, Santa Barbara, CA, makes the GeneChip® which is a flat plate array, but Illumina, San Diego, CA, makes a fiber optic bundle in which the active probes are attached to microbeads that reside in small wells in the end of the fiber. Yershov et al. (1996) reported an oligonucleotide array incorporated into gel elements patterned onto a glass plate with a 1.5 cm^2 chip containing 20 – 30 K elements. The one common factor that all of these technologies have is that the substrates are usually inorganic glass for two reasons: primarily the optical quality of the glass because the probes are usually interrogated using light, but also to remove the captured DNA, heating to nearly 100℃ is required. The glass can be made to accommodate a rather high density of oligonucleotide probes by various methods, predominantly due to the flexibility of composition and manufacturing. Selective oligonucleotides can be patterned onto the substrate, for example, via a photolithographic technique as in the case of Affymetrix. With only 4 nucleotides, one could achieve 256 possible permutations (Goldstein, 1998). Illumina (Wells, 1998; Czarnik, 1998) has targeted SNP's (single nucleotide polymorphisms) by attaching one type of oligonucleotide to a microscopic bead and each individual bead resides in a separate well at the end of a massive fiber bundle. Each bead, therefore, can act as a probe for a particular sequence. A variety of materials have been used for the beads, but the fiber bundle is always high optical quality glass.

The specificity and binding capabilities of glass have not escaped the notice of experimenters dealing in the isolation and purification of DNA. Most commercially available DNA purification products use glass or other silicate-based materials as the substrate to which the DNA absorbs to extract it from other molecular mixtures (Goldstein, 1998). However, the efficiency of these materials for binding the DNA is rather low; in fact, the efficiency has not really improved significantly since the original reports of glass as a binding agent (Vogelstein and Gillespie, 1979), and the binding capacities are of the order of 2 – 10 μg/mg. There appears to be little evidence in the literature of attempts to optimize the binding in any way, although there may have been unpublished proprietary research. Most of the data associated with the DNA purification kits do not allude to chemistry or surface treatment or manufacturing, which suggests that there may be room for materials optimization.

Considering the amount of technology that is based upon attachment of proteins and other biomolecules to glass, very little is known about the interactions that govern this. There is evidence through cell interactions that commercial glasses, well within the ranges of compositions used in biotechnology, behave very differently and that the assumption that "a glass, is a glass, is a glass" may not be correct (Jedlicka and Clare, 1999). It was also noticed (Chan, 1999) that in making porous glass substrates from a Bioglass®

composition, human breast cancer cells were preferentially attracted to the pellets, whereas human colon cancer cells were not. This is further evidence that the surface of Bioglass-like compositions present a bone-like structure on their surface since breast cancer cells tend to metastasize to bone.

3.3.3 Optical Properties

One of the many reasons that glass is favored in a majority of applications is its optical properties. The usually electrically insulating nature of glass (although some glasses are semiconductors, and metallic glasses are conductors), coupled with the unique structure of glass (absence of pores, absence of grain boundaries, homogeneous structure), gives most glass transparency in the visible region of the spectrum and to a certain extent beyond. The transparency of glass is limited at the short wavelength end by the electronic band gap of the glass. Typically, the more ionic glasses (e. g. halides) have higher UV transparency than covalent glasses because they are more electrically insulating, although high-purity vitreous silica is often used for UV transmission. UV transparency becomes important for some sterilization procedures but also for laser surgery. High energy light such as UV induces a photochemical effect that can lead to bond breakage without excessive heating. High absorption coefficients for UV light coupled with short wavelengths can lead to very high resolution. This is particularly important for eye surgery such as PRK (photorefractive keratectomy) which is laser sculpting, where UV light coupled with a negative mask can be used to sculpt the outer cornea to adjust the focus to the retina, LASIK (laser assisted in-situ keratomileusis) which is similar but involves sculpting the inner cornea, or radial keratotomy where precise relief cuts are made in the eye lens to relax the shape. Optical fiber delivery greatly facilitates the process, and glasses that have a high UV transparency are important.

The fundamental vibrations of the atoms in the glass network limit the transparency of a glass at the long wavelength end. In practically all inorganic glasses, the structural bonds possess a permanent dipole moment that can lead to infrared absorption. Infrared absorption is governed by the force constant of the bonds and the mass of the atoms, so higher IR transparency can usually be achieved by moving toward materials with heavier atoms and lower force constants. The chief applications in the infrared are in surgical procedures and IR sensing.

For surgery, it is best to couple the light directly into the tissue which needs to be cut or ablated. Although most tissues contain more than 70% water, the absorption of light by tissue varies considerably with both tissue type and general tissue health. For example, water has a peak around 3 μm, remains high out to 10 μm, but tissues that contain chromophores like hemoglobin show high absorption in the visible region. Maximization of the absorption by a tissue reduces peripheral damage and both forward and back-scattering of the laser light. For photothermal heating, typically the further one goes into the IR, the better the cutting resolution. Again, laser light delivery can be an issue particularly as many of the lasers used in surgery are fairly large and quite cumbersome. It is probably a larger issue in the case of the IR because silica, the material for which the technology of optical fiber fabrication is best known, is limited to wavelengths below about 3 μm. This basically allows the Ho: YAG laser light but the transmission of Er: YAG laser light centered around 3 μm usually requires a special fiber such as fluoride glass. CO_2 laser light which is probably the best for IR surgery is inaccessible to all of these glasses and hence can only currently be delivered by hollow fiber or single crystal fibers both of which have their own associated problems. There have been some studies advocating chalcogenide glasses for this application.

Chalcogenide glasses are made from arsenic, antimony, tellurium, or germanium mixed with sulfur or selenium. These glasses are dark red to black in color but transmit quite far into the infrared provided they are made with reasonable purity (King et al., 1995). However, other issues such as self-focusing have to be solved before wide application of these glasses to the transmission of CO_2 laser light. However, when not dealing with high power densities, chalcogenide glass fibers have found applications in IR sensing. Chalcogenide glasses have a very wide IR transmitting optical window, which makes them applicable to this field. In the same way that the absorption of IR light can help to heat tissue the characteristic absorption of light by tissue and its subsequent luminescence can reveal much about its general well-being.

Throughout the spectrum from UV through IR, the absorption of light and the subsequent luminescence can be used to assess the status and health of tissue. Glass fiber is usually used as the delivery and acceptance device for light to, and from, the tissue of interest. Clearly, the glass that is chosen for the optical fiber delivery mechanism must have a transmission window suitable for that region, but moreover it is important, particularly if dealing with low light levels, that there be no impurities in the glass which may absorb or luminesce at the point where one expects to see changes. Telecommunications fibers can be made by a chemical vapor deposition route to have lower than ppb of impurity. The main culprit ions interfering with optical signals are transition metal or rare earth ions. While rare earth contamination is relatively uncommon, transition metals are a common contaminant in many of the raw materials used to make even good optical glass. Transition metals absorb and, in many cases fluoresce, so if one wants a glass, say, for diagnostics, either for a device like the GeneChip® or an optical fiber device, one must avoid such contaminants. Contamination in a long path length device involving optical fiber is probably more serious than in a short path length device such as a 2-D array, but purification technologies in optical fibers are quite well developed due to the communications industry.

In some cases, light is used to activate a reaction as in photodynamic therapy. This technique involves a chromophore that is selectively retained in unhealthy tissue which, when activated by light, can cause the release of a local agent, allowing selective treatment of unhealthy tissue. If glass is used as a delivery device for this type of therapy, then one must ensure that it is as transparent as possible in the region of the activating light. General excellent transmission over the visible region of the spectrum is also what is required for endoscopes, so that a reliable image can be transmitted. Glass is the material of choice again for its optical quality but also because making large fiber optic endoscopic bundles is very successful with glass. To make an image, transmission device fibers with a large core and small cladding are made, packed together (called "multi's") in as close-packed array as is allowed by the geometry, and repeatedly pulled and stacked until the fibers are of the order of a few microns in diameter with tens of thousands of fibers in one bundle. At this point one can achieve very high resolution of an image. Endoscopes can be rigid or flexible, since optical fibers are inherently very flexible as well as strong. Excellent transmission over a broad band is an issue with these fibers since one is effectively looking through a long path length of glass. If one looks at window glass from the side, one sees a quite intense green due to contamination by iron. Window glass, therefore, would not be suitable for this application. High quality optical glass without a high impurity or defect concentration is required.

The chemical and structural flexibility of glasses means that they do not have to be only passive optical sensors, as previously described. Glasses can also be made active optical sensors of biological activity by encapsulating biomolecules (usually proteins) into a sol-gel glass matrix. In the sol-gel method, mentioned previously, a chemical reaction produces the glass into which proteins may be bound and still remain active (Johnson and Whateley, 1971). This happens by the hydrolysis and condensation of an organosilicate. Usually, a cosolvent is required to enable the reaction between the organosilicate and water, which are immiscible, and the most common cosolvent is ethanol. Many proteins are sensitive to ethanol (Cabral and Kennedy, 1993) and can become denatured. Additionally, the presence of a cosolvent causes higher porosity (Zarzycki, 1992). It is possible, however, using external agitation such as ultrasonic waves, to enable enough mixing to allow the reaction not only to proceed, but to proceed a great deal faster than some of the other techniques. This technique also allows manipulation of the pH to levels that can be supported physiologically for the addition of a protein (Hall, 2000). Some optically active biomolecules can be incorporated into a sol-gel glass in this way and can potentially lead to active sensors by being coated onto optical components. Provided the gel is designed correctly, the porosity should be such that there is enough access to the biomolecule to trigger optical activity.

One of the oldest applications of glasses and glass-ceramics has been in dentistry. Probably the most important feature of a dental material these days is aesthetics. The translucency and color of natural teeth are imitated most closely by carefully constructed glass-ceramics. Until recently, these have generally been carefully enameled dental alloys but in the pursuit of more attractive dental restorations, there has been a move toward all-ceramic restorations where the outside layers are still laminated glass-ceramics, but the interior is also ceramic. These all-ceramic restorations have issues to be solved regarding strength and reliability (Pospiech et al., 2000; Ludwig et al., 2000), but they are undoubtedly an improvement in terms of the optical quality of the restoration.

3.3.4 Mechanical Properties

Glasses are not known for their superior mechanical properties. In fact, bioactive materials such as Bioglass® would probably be much more widely applied if they had better mechanical properties, really referring to fracture toughness. Some of the glass-ceramic material alternatives, however, such as Ceravital®, Cerabone®, and Bioverit®, which are glass-ceramics containing crystalline apatite, an apatite – wollastonite glass-ceramic, and a glass-ceramic containing apatite and fluorophlogopite, respectively (Kokubo, 2000), are much tougher and have some demonstrated bone-bonding capabilities to different extents. Glass-ceramics have the advantage that the presence of the second phase can serve to block the cracks that cause catastrophic failure in glasses. In addition, dependent upon the mechanism of growth, the morphology and aspect ratio of the crystalline phases can serve to enhance the strength in ways similar to that of the mineral portion of bone itself (Likitvanichkul and LaCourse, 1998).

Of course, similar strength and fracture toughness can be achieved by using optical fiber reinforcement for other materials, as alluded to in the chemical durability section. Glass fiber is a very high strength material especially when it first emerges from the fiber drawing process when the surfaces are flawless. In the case of a composite such as a bioresorbable bone fixation device, one is relying on the dissipation of the crack energy at the interface between the fiber and the matrix phase to increase the strength of the material. Not only glass fiber but also glass particulates have been used for many years in glass ionomer cements in dentistry.

It was mentioned in Sect. 3.3.3 that dental restorations are moving away from glass and glass-ceramic coatings on dental alloys toward glass and glass-ceramic coatings on ceramic bodies. The reason is aesthetics. Exposure at the gum line can result in a dark line if a dental alloy is used, whereas this does not happen with an all-ceramic restoration. In addition, realistic teeth have a translucency to them that is not possible with an alloy body. Most of the problems lie in the strength of the all-ceramic restorations (Ludwig et al., 2000; Pospiech et al., 2000). Anterior chewing forces are at about 180 N (Ludwig et al., 2000), and although ceramics can be made that exceed that force in terms of their fracture resistance, the scatter in the strength data for ceramic materials can cause concerns in terms of reliability. Nevertheless, there have been numerous improvements in the reliability of all-ceramic restorations, and it is anticipated that improvements will continue causing these restorations to command a much larger portion of the market.

Whether for use as an enamel for a dental alloy or as a coating for an all-ceramic restoration, one mechanical property that is quite important is hardness. The glasses and glass-ceramics used to mimic natural tooth dentin tend to be much harder materials than natural dentin. As a result, the restoration can cause undue wear on the patient's natural teeth. It is essential therefore that the restoration be as smooth as possible to minimize wear. Glasses and glass-ceramics tend to be amenable to polishing, have very smooth surfaces after firing, and hence are not unreasonable in terms of wear; however, a compromise with perhaps a composite material might be a way of alleviating this problem in the future.

3.3.5 Thermal, Electric, and Magnetic Properties

These properties of glasses really only have isolated applicability. Thermal properties are important in the manufacture but do not raise too many concerns during applications, although the ability to autoclave and dry heat treat glass is one of the reasons that it is such a versatile biotechnological material. The mouth is probably the location in which a glass biomaterial will experience the most thermal cycling. In dentistry, there exists mild concern for thermal shock resistance in cold weather, but it is usually not an issue. Thermal conductivity of a dental restoration may contribute to discomfort, but typically thermal expansion coefficients over the temperature range in the mouth are also not an issue.

Electrical properties are not usually key in most biotechnological and biomedical applications of glass; glasses are, for the most part, insulators. Nevertheless, the pacemaker for the human heart involves glass seals for the electrical feed-throughs because of the superior chemical durability of glass and the ability to compositionally design a glass to seal to metal with a tailored thermal expansion coefficient.

The magnetic properties of glasses and glass-ceramics do not usually play a prominent role in medical and biotechnological applications. However, some glass-ceramic compositions discovered by Luderer et al. (1983) had a ferrite phase crystallized out that exhibited a magnetic hysteresis heating effect when exposed to a moderate [500 Oe (1 Oe ≈ 79.5775 A/m) at 10 kHz] alternating magnetic field. The application for which this material was designed was the hyperthermic treatment of cancer. It is reported that temperatures in the range 42℃ – 45℃ cause the destruction of malignant cells (Cavaliere et al., 1967; Overgaard and Overgaard, 1972; Overgaard, 1977). This does depend upon several other factors related to the character of the tumor (Luderer et al., 1983). Raising the whole body temperature is both dangerous and difficult, but localized heating can be employed to destroy malignant tissue.

Borrelli et al. (1984) demonstrated this concept in rodent models that had been implanted with a Meth-A sarcoma. The rodents contained a control group (no treatment), a group that had ceramic injected subcutaneously close to the tumor, and a group that was subjected to a 10 kHz 500 Oe field. A final group was subjected to both ceramic and one 2-hour field treatment. In the last group, an approximately 50% tumor regrowth delay and a 12% permanent control was observed. Bearing in mind that this was the result of one magnetic field treatment only, these were impressive results. Magnetic hysteresis heating has also been employed by Ebisawa et al. (1993) in a bioactive ceramic material to treat cancer of the bone.

3.4 Summary

Inorganic glass has a ubiquitous role in biotechnology and has excellent potential as a biomedical material, and yet it is not a material that immediately springs to mind. Glass is the material in the background, waiting in the wings for its entrance to the stage. Glass is that character actor you see in movies all the time, but you never know his or her name because they always play something different and show such versatility that you just do not realize it is the same person. Glass could take on much larger roles, given the opportunity and can even be far more versatile in the roles that it already has with more research and nurturing. The future of glass in telecommunications is clear, and I believe that glass has its niche in biotechnology and biomedicine for all the same reasons.

References

Borrelli, N. F., A. A. Luderer, J. N. Panzarino. Phys. Med. Biol.. 29:487 - 494 (1984)

Cabral, J. M. S., J. F. Kennedy. In: M. N. Gupta, ed. *Thermostability of Enzymes*, Cha. p. 10. Springer-Verlag, NY (1993)

Cavaliere, R., E. C. Ciocatto, B. C. Giovanella, C. Heidelburger, R. O. Johnson, M. Margottini, B. Mondovi, B. G. Moricca, A. Rossi-Fanelli. Cancer. 20: 1351 - 1381 (1967)

Chan, M. M. S. Thesis. Alfred University, unpublished (1999)

Chee, M., R. Yang, E. Hubbell, A. Berno, X. C. Huang, D. Stern, J. Winkler, D. J. Lockhart, M. S. Morris, S. P. Fodor. Science. 274:610 - 614 (1996)

Choueka, J., J. L. Charvet, H. Alexander, Y. H. Oh, G. Joseph, N. Blumenthal, W. LaCourse. J. Biomedi. Materi. Res. 29:1309 - 1315 (1995)

Czarnik, A. W.. Modern Drug Discoverty. 49 - 55: November/December (1998)

Day, D. E. Bull. Am Ceram. Soc.. 74:64 - 68 (1995a)

Day, D. E. *Proc. 17th Int. Congr. Glass.* vol 1:243 - 251 (1995b)

Ebisawa, Y., T. Kokubo, K. Ohura, T. Yamamuro. Mater. Med.. 3:225 - 232 (1993)

Goldstein, A. H. Report to NSF Industry/University Center for Biosurfaces (1998)

Hall, M. M. MS Thesis. Alfred University unpublished (2000)

Hench, L. L., J. K. West. Life Chem. Rep. 13:187 - 241 (1996)

Hench, L. L., J. Wilson. In: E. D. Clarke, J. Simmons, and D. Folz, ed. *Ceramic Transactions*. p. 101, American Ceramic Society, OH (1999)
Hubert, F. Rivista della Stazione Sperimentale del Vetro. 6:29 – 35 (1986)
Jedlicka, A. B., A. G. Clare. In: E. D., Clarke, J. Simmons, and D. Folz, eds. *Ceramic Transactions*. p. 101, American Ceramic Society, OH (1999)
Johnson, P., T. L. Whateley. J. Colloid Interface Sci. 37:557 – 563 (1971)
King, W. A., A. G. Clare, W. C. LaCourse. J. Non-Cryst. Solids. 181:231 – 237 (1995)
Kokubo, T., H. Kim, M. Kawashita, H. Takadama, T. Miyazaki, M. Uchida, T. Nakamura. Glasstech. Ber. 73C1:247 – 255 (2000)
LaCourse, W. C. Private Communication. unpublished (1996)
Likitvanichkul, S., W. C. LaCourse. J. Mater. Sci. 33:5901 – 5904 (1998)
Lin, S. T., S. L. Krebs, S. Kadiyala, K. W. Leong, W. C. LaCourse, B. Kumar. Biomaterials. 15:1057 – 1061 (1994)
Luderer, A. A., N. F. Borrelli, J. N. Panzarino, G. R. Mansfield, D. M. Hess, J. L. Brown, E. H. Barnett. Radiat. Res. 94:190 – 198 (1983)
Ludwig, K., S. Kubick, S. Klopfer. Glasstech. Ber. 73C1:293 – 302 (2000)
Marcolongo, M., P. Ducheyne, W. LaCourse. J. Biomedi. Mater. Res. (1996)
McIntyre, A. D. E. Day. Phys. Chem Glasses. 39:29 – 35 (1998a)
McIntyre, A. D. E. Day. *Proc. XVIII Int. Congress Glass* CD ROM (1998b)
Muller-Mai, C. M., C. Voigt, R. E. Baier, U. M. Gross. Cells and Mater. 2:309 – 327 (1992)
Overgaard, K., J. Overgaard. Eur. J. Cancer. 8:65 – 78 (1972)
Overgaard, J. Cancer. 39:2637 – 2646 (1977)
Pospiech, P., St. Kistler, C. Frasch. Glasstech. Ber. 73C1:302 – 310 (2000)
Scholze H. Glasstech. Ber. 63:141 – 147 (1990)
Vogelstein, B., D. Gillespie. Proc. National Academy of Science. USA, 76:615 – 619 (1979)
Wells, W. Chemi. and Biol. 5 (1998)
Yershov, G., V. Barsky, A. Belgovskiy, E. Kirillov, E. Kreindlin, I. Ivanov, S. Parinov, D. Guschin, A. Drobishev, S. Dubiley, A. Mirzabekov. Proc. National Academy of Science. USA, 93:4913 – 4918 (1996)
Zarzycki, J. *Ultrastructure Processing of Advanced Ceramics*. In: D. R. Uhlmann and D. R. Ulrich, eds. John Wiley and Sons, NY (1992)

4 Biocomposite Materials for Biotechnology

Toshihiro Kasuga, Hideo Hosono, Yoshihiro Abe

Properties of surfaces and near-surface regions of materials are greatly enhanced in porous bodies. Porous glasses and glass-ceramics utilizing spinodal-type phase separation in glass have great advantages over porous ceramics prepared by a conventional powder-sintering method. Since the first porous phosphate glass-ceramic was reported in 1989, a series of glass-ceramics in titanium phosphates was developed. The present chapter describes the synthesis and characteristic properties of porous glass-ceramics for applications to the biotechnology field such as supports for immobilization of enzymes and bacteriostatic materials.

Bulk porous glass-ceramics with a functional crystal skeleton were prepared by controlled crystallization of titanium phosphate glasses and subsequent acid leaching. A mean pore diameter and porosity of 15 – 200 nm and 40vol% – 60vol% were achieved. Porous $CaTi_4(PO_4)_6$ glass-ceramics have excellent properties as supports for immobilization of enzymes. Conspicuous bacteriostatic activities were shown for porous glass-ceramics with the surface phase consisting predominantly of $AgTi_2(PO_4)_3$ crystal and the interior phase of $LiTi_2(PO_4)_3$ crystal and those with a skeleton of $CuTi_4(PO_4)_6$ crystal. Porous glass-ceramics with an integrated skeleton of $AgTi_2(PO_4)_3$ and $Ti(HPO_4)_2 \cdot 2H_2O$ crystals can also be synthesized utilizing a finding that the $LiTi_2(PO_4)_3$ crystal with a three-dimensional network structure of TiO_6 and PO_4 is transformed into $Ti(HPO_4)_2 \cdot 2H_2O$ with a two-dimensional layered structure in acid solution: the materials showed various functions such as bacteriostatic activities, adsorption of ammonia gas, and intercalation of polar molecules. Porous glass-ceramics with a skeleton of titanium phosphate crystal are excellent materials that have the combined advantages of porosity, ease of fabrication, and crystal functionality.

4.1 Introduction

Since the thermal, chemical, and mechanical properties of ceramics are much better than those of metals and organic polymers, porous ceramics are used in various engineering fields such as catalytic supports (Aso et al., 1979), membranes (Yazawa and Tanaka, 1993), and sensors (Uo et al., 1992). Novel uses, such as immobilized enzymes (Weetal, 1969a; Messing, 1977) and packing materials (Haller, 1965) in chromatography, are also anticipated in biotechnology.

In some glasses, spinodal-type phase separation occurs. Both of the separated phases in the glasses are continuous (Rawson, 1967). Microporous glasses and glass-ceramics (ceramics prepared by controlled crystallization of glass) utilizing spinodal-type phase separation in glass have great advantages over conventional porous ceramics prepared by sintering of powdery materials because they have well-controlled pores (Volf, 1961; Beall and Duke, 1969). A schematic drawing of the preparation procedure of porous glasses and glass-ceramics utilizing spinodal-type phase separation is shown in Fig. 4.1. Research and development on porous glasses and glass-ceramics, however, have mainly been focused on silica-based systems (Res et al., 1984; Rabinovich et al., 1981; Schmid et al., 1988) following the invention of porous silica glass (Vycor®) (Hood and Nordberg, 1934; Hammel, 1974), which is obtained from $Na_2O - B_2O_3 - SiO_2$ glass. If porous glass-ceramics with a skeleton of crystals with active functions can be fabricated, they will have the combined advantages of porous bodies (pores, large surface area), glass-ceramics (ease of fabrication), and crystal skeletons (functionality).

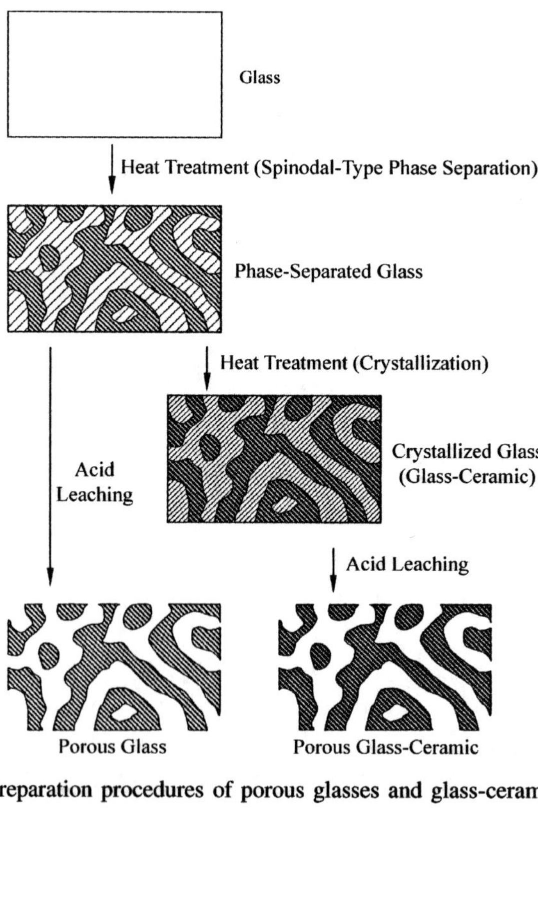

Figure 4.1 Preparation procedures of porous glasses and glass-ceramics (schematic)

The authors have been studying phosphate glasses and glass-ceramics for application to biorelated fields. Some phosphates with functionality are known. The authors' group reported (Hosono et al., 1989) that porous glass-ceramics with a skeleton of $CaTi_4(PO_4)_6$ crystal (Roy et al., 1984) can be prepared by controlled crystallization of glasses in the system $CaO-TiO_2-P_2O_5$ and subsequent acid leaching of the resulting dense glass-ceramics composed of $CaTi_4(PO_4)_6$ and β-$Ca_3(PO_4)_2$ phases which interlock with each other. The porosity is ~50vol%, and the median pore diameter and specific surface area are approximately 40 nm and 50 m^2, respectively. Since the development of the first porous glass-ceramic based on phosphates, the authors have prepared various porous glass-ceramics with a skeleton of functional titanium phosphate crystals.

4.2 High-Performance Titanium Phosphates

Two types of high performance crystals are known in the $TiO_2-P_2O_5$ system (Kanazawa, 1990). One is the so-called Nasicon-type (Goodenough et al., 1976) crystal such as $RTi_2(PO_4)_3$ with rhombohedral $R\bar{3}C$, as shown in Fig. 4.2(a). PO_4 tetrahedra in the structure share their corners with TiO_6 octahedra to form a three-dimensional network (Hagman and Kierkegaard, 1968). With Li^+ ions (four possible sites per formula unit) located in two different sites of the conduction channels, $LiTi_2(PO_4)_3$ and their solid solutions have attracted considerable attention as chemically stable, fast Liconductors (Subramanian et al., 1986; Aono et al., 1989) for the electrode materials of a secondary battery.

The other is a zirconium phosphate type crystal, $Ti(HPO_4)_2 \cdot nH_2O$. This type of crystal has a two-dimensional layered structure (Clearfield and Smith,

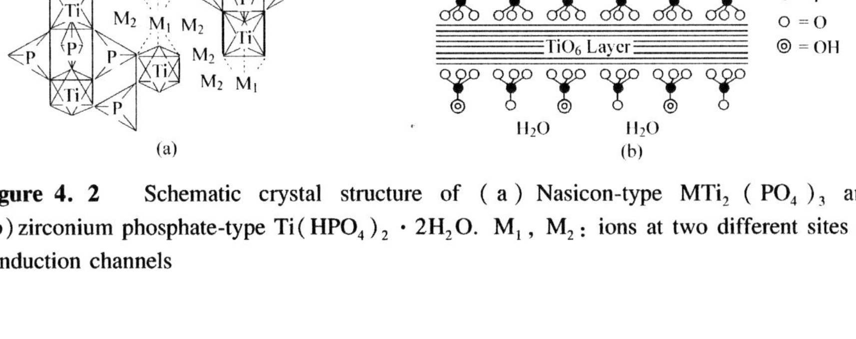

Figure 4. 2 Schematic crystal structure of (a) Nasicon-type $MTi_2(PO_4)_3$ and (b) zirconium phosphate-type $Ti(HPO_4)_2 \cdot 2H_2O$. M_1, M_2: ions at two different sites of conduction channels

1969) as shown in Fig. 4.2(b); Ti^{4+} ions lie very nearly in a plane and are bridged by phosphate groups, which are situated alternatively above and below the metal planes. Three oxygen atoms of each PO_4 group are bonded to three different Ti^{4+}. The fourth oxygen atom of the PO_4 group has an exchangeable proton. The proton and layered structure are the structural origin of excellent ion-exchange properties (Clearfield, 1982) and permit intercalation of polar organic molecules (Yamanaka, 1989).

4.3 Preparation Procedures of Porous Glass-Ceramics

Neither the chemical composition of the Nasicon-type crystal nor the zirconium phosphate-type crystal can be vitrified by conventional quenching techniques (Kishioka,1978). However, glass formation occurs in the pseudobinary system $RTi_2(PO_4)_3$ (R = Li, Na, 1/2Ca) – $Ca_3(PO_4)_2$, but neither binary forms a stable glass. This finding is consistent with an empirical rule that a composition near that of a eutectic point has a high tendency to make a stable glass.

A mixture of starting materials such as metal carbonates, TiO_2, and H_3PO_4 (85% liquid) were placed with water in a Teflon® beaker and stirred well to make a slurry. The slurries were dried at about 200℃. The resulting dried products were melted in platinum crucibles at 1300℃ for 1 h in air. Corrosion of platinum crucibles by the melt was not perceptible. The melts were poured into a carbon mold and subsequently annealed. The resulting glasses were transparent but deep violet in color. This color is due to absorption of Ti^{3+} (~0.1% of total Ti) in the glasses (Hosono et al., 1989). The glasses were cut into rectangular plates (~ 15 mm × 15 mm × 5 mm). Glass plates were crystallized by a two step heat treatment. First, the specimens were heated to the glass transition temperature (Tg) + 20℃ (Tg is determined by differential thermal analysis (DTA) employing a heating rate of 10 K/min) for 20 h to promote nucleation. After this treatment, the color of the glass changes from deep violet to almost colorless due to the oxidation of Ti^{3+} by residual hydroxyl groups in the glass (Hosono et al., 1989). Subsequently, the specimens were heated to a temperature of Tc – 40℃, where Tc is the peak crystallization temperature on the DTA curve, for 12 h to crystallize them. The resulting dense glass-ceramics consisting of two crystalline phases, $RTi_2(PO_4)_3$ and β-$Ca_3(PO_4)_2$, which interlock with each other, were soaked in 1-N aqueous HCl solution and kept for 24 h. Since the solubility of β-$Ca_3(PO_4)_2$ (denoted by TCP), which is known to be a bioactive material like hydroxyapatite (Monma and Kanazawa, 1976; Doremus, 1992), in aqueous solution increases drastically as the pH of the leaching solution is lowered (Monma and Kanazawa, 1976; Chow, 1991), the acids selectively dissolve β-$Ca_3(PO_4)_2$, leaving a skeleton of $RTi_2(PO_4)_3$.

Figure 4.3 shows some examples of powder X-ray diffraction (XRD) patterns of glass-ceramics with acid leaching. The following are evident from the figure: (a) The dense glass-ceramics are composed of $RTi_2(PO_4)_3$(R = Li, Na,

1/2 Ca) and β-$Ca_3(PO_4)_2$. (b) The acid dissolves TCP selectively. (c) $LiTi_2(PO_4)_3$ (denoted by LTP) is converted into $Ti(HPO_4)_2 \cdot 2H_2O$ (denoted by THP) in acids at around 100℃. This conversion does not occur in neutral and alkaline aqueous solutions even at 100℃. It is noteworthy that LTP with a three-dimensional network structure is converted into THP with a two-dimensional layered structure under such mild conditions. This reaction is specific to LTP; $NaTi_2(PO_4)_3$ (denoted by NTP) and $CaTi_4(PO_4)_6$ (denoted by CTP) are stable under the same conditions (Hosono and Abe, 1992a). Figure 4.4 shows the pore size distribution and microstructure of porous glass-ceramics (Hosono and Abe, 1992b) with a skeleton of $Li_{1+x}Ti_{2-x}Al_x(PO_4)_3$ ($x = 0.4$), which is a chemically stable fast Liconductor. As a

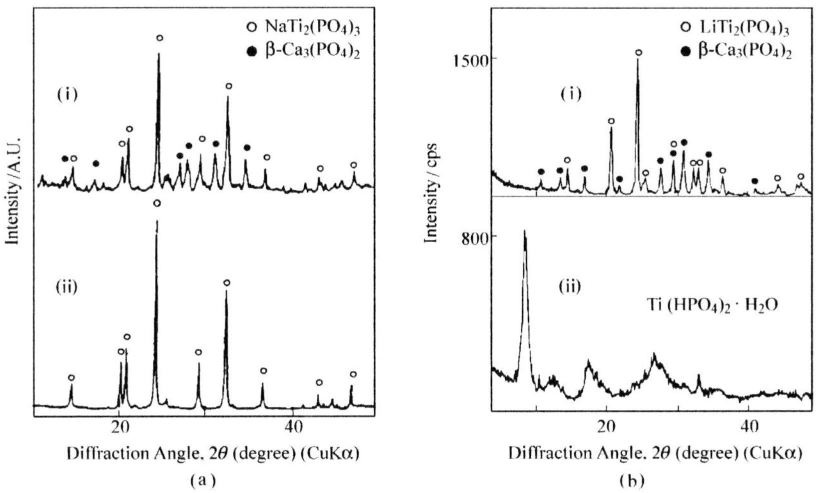

Figure 4.3 Changes in the powder XRD pattern of glass-ceramics with acid leaching. Leaching was done in 1 *N*HCl at 100℃: (a) $NaTi_2(PO_4)_3 - Ca_3(PO_4)_2$ system and (b) $LiTi_2(PO_4)_3 - Ca_3(PO_4)_2$ system. (i) Before leaching and (ii) after leaching

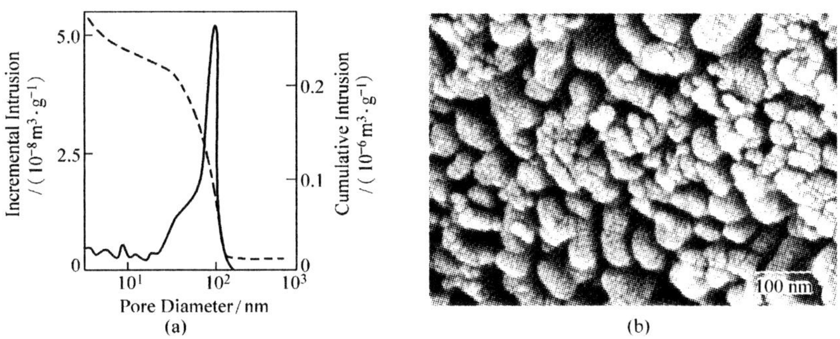

Figure 4.4 Porous glass-ceramics with a skeleton of $NaTi_2(PO_4)_3$. (a) Pore size distribution curve and (b) SEM photo of fracture face

first approximation, the distribution is log Gaussian in shape, and the median pore diameter is approximately 0.2 mm. This diameter is the maximum among the porous phosphate glass-ceramics. The diameter of the porous glass-ceramics is in the range of 20 – 60 nm (Hosono and Abe, 1993).

4.4 Porous Glass-Ceramics for Immobilization of Enzymes

Enzymes are scarce biological catalysts with selectivity specific to substrate the and very high efficiency. The application of enzymes is limited by their scarcity and cost; therefore, immobilization is indispensable for practical use (Weetal, 1969a). The optimum pore diameter for immobilizing an enzyme is approximately twice the major axis of the unit cell of the enzyme (Messing, 1977). In most cases, the optimum pore diameter corresponds to 30 – 60 nm.

Figure 4.5 shows the pore size distribution of porous $CaTi_4(PO_4)_6$ glass-ceramic. The distribution is almost log Gaussian in shape, and the width is rather sharp (Hosono et al., 1992). The median pore diameter is estimated to be 35 nm, which is in an appropriate range for immobilization of enzymes. The BET specific surface area is ~ 70 m^2/g. Figure 4.6 shows a photograph of the

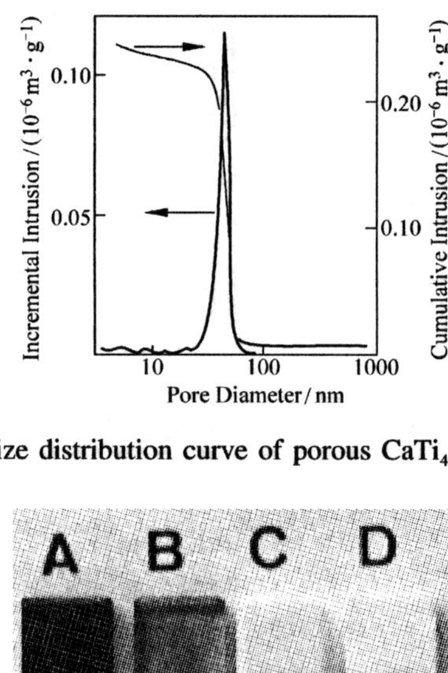

Figure 4.5 Pore size distribution curve of porous $CaTi_4(PO_4)_6$ glass-ceramic

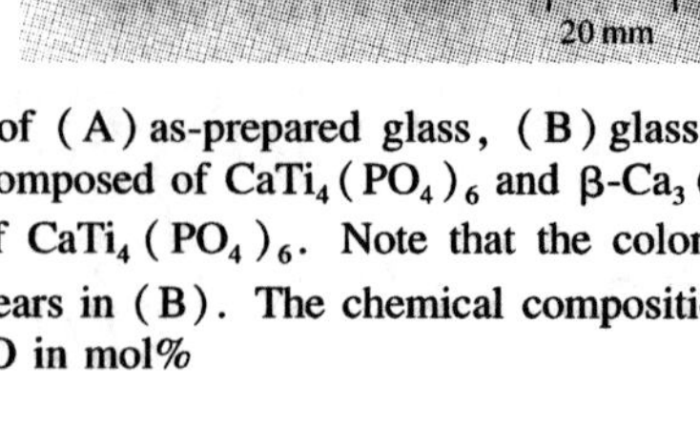

Figure 4.6 Photograph of (A) as-prepared glass, (B) glass after nucleation treatment, (C) dense glass-ceramic composed of $CaTi_4(PO_4)_6$ and β-$Ca_3(PO_4)_2$, (D) porous glass-ceramic with a skeleton of $CaTi_4(PO_4)_6$. Note that the color in (A), which is due to residual Ti^{3+} ions, disappears in (B). The chemical composition of the glass is $45CaO \cdot 25TiO_2 \cdot 30P_2O_5 + 2Na_2O$ in mol%

as-prepared glass (A), the glass (B) after nucleation, the dense glass-ceramic (C) before leaching, and the porous glass-ceramic (D) prepared by leaching. It is obvious from the photo that no cracking, shrinkage, or deformation occurs in the course of crystallization and subsequent leaching.

4.4.1 Immobilization of Enzymes and Measurement of Activity

Porous $CaTi_4(PO_4)_6$ glass-ceramics (denoted by CTP) and controlled pore SiO_2 glass (denoted by CPG), which were purchased from a supplier, were examined as carriers. The materials were crushed into beads. The beads were sieved to obtain narrow sieve fractions with diameters from 1 – 2 mm. Table 4.1 summarizes the properties of the two carriers examined. The pore sizes and surface areas are similar to each other. As shown in Fig. 4.7, a remarkable difference is seen in the alkaline durability (Hosono et al., 1991; Suzuki et al., 1991). CPG is stable in acidic and neutral solutions but is easily dissolved in solutions with pH > 9. On the other hand, CTP is stable to either acidic or alkaline solutions because it contains a large amount of TiO_2.

Table 4.1 Properties of enzyme carriers examined

	CTP (Porous $CaTi_4(PO_4)_6$ glass-ceramic)	CPG® (Porous SiO_2 glass)
BET surface area/m^2/g	70	84
Median pore diameter/nm	35	32
Pore volume/cm^3/g	0.46	0.64

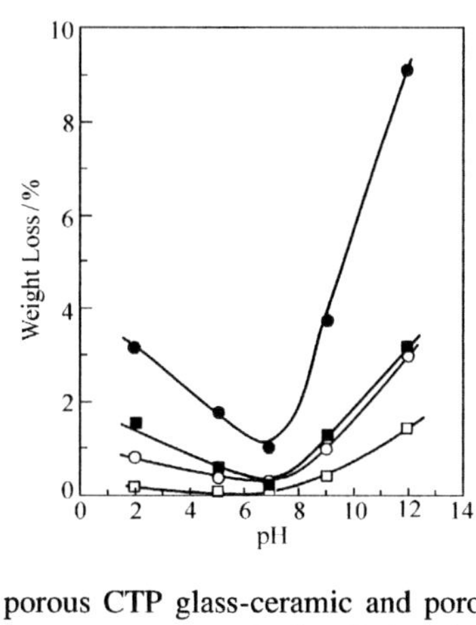

Figure 4.7 Durability of porous CTP glass-ceramic and porous CPG glass in various pH solutions at 37℃. Weight loss was measured after 7 d and 14 d soaking, respectively. (○): CPG and (□): CTP at 7 d; (●): CPG and (■): CTP at 14 d

Enzymes were cross-linked to aminosilanizated CTP and CPG carriers by glutaraldehyde, as reported by Weetall (Weetall, 1969b). Beads of the materials were refluxed for 6 h in toluene containing 2% γ-aminopropyl triethoxysilane at 110℃ to silanizate them. After the silanizated beads were dried in an oven at 100℃, the dried beads were packed into a glass column of 5 mm diameter. The packed beads were reacted for 24 h at 4℃ with 2.5% glutaraldehyde solution buffered to pH 7.5 by circulating the buffer with a peristaltic pump. Then, an enzyme solution buffered to pH 7.5 was fed to the packed beads and circulated through the column for 24 h at 4℃.

The enzyme reactor shown in Fig. 4.8 was designed to analyze the retained activities of immobilized derivatives at a constant temperature and feed rate (Suzuki et al., 1991).

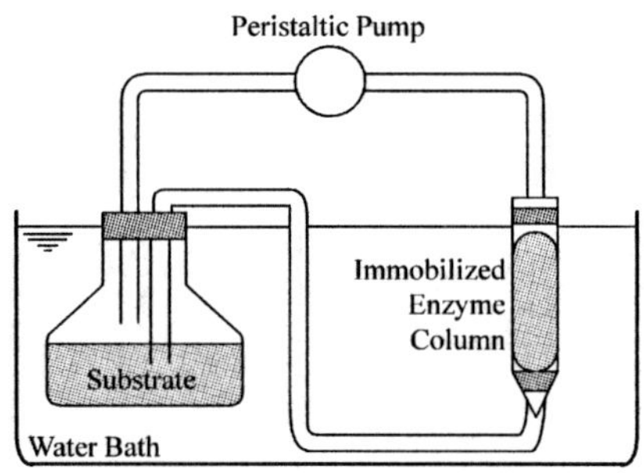

Figure 4.8 Setup to measure the activity of the immobilized enzyme

4.4.2 Immobilization of β-Galactosidase

β-galactosidase from Escherichia coli, whose activity is maximal in a neutral pH condition, had an activity of 161 U/mg where one U is the amount of enzyme that hydrolyzes 1 μmol of O – nitrophenyl-β-D-galactopyranoside (ONPG)/min at 25℃ and pH 7.5. The activity was calculated from the rate of hydrolysis of ONPG in phosphate buffer solution, pH 7.5, at 25℃. The appearance of O – nitrophenol was followed by measurement of the extinction at 420 nm of the reaction mixture.

Figure 4.9 shows the time courses of the remaining activities. The reduction curves at 37℃ show that the CTP derivative was very stable. In contrast, the CPG derivative lost its activity more quickly since it is somewhat soluble in a neutral solution.

The result at 50℃ shows that the activities of the CTP and CPG derivatives were rapidly lost during the initial 50 min. The amount of protein released in the effluent from the CPG derivative at 50 min reached 0.24 mg, which was 48% of the initially attached protein on the carrier. On the other hand, 0.11 mg of protein was detected in the effluent of the CTP derivative. These results show that the first rapid reduction depended on the release of insufficiently attached

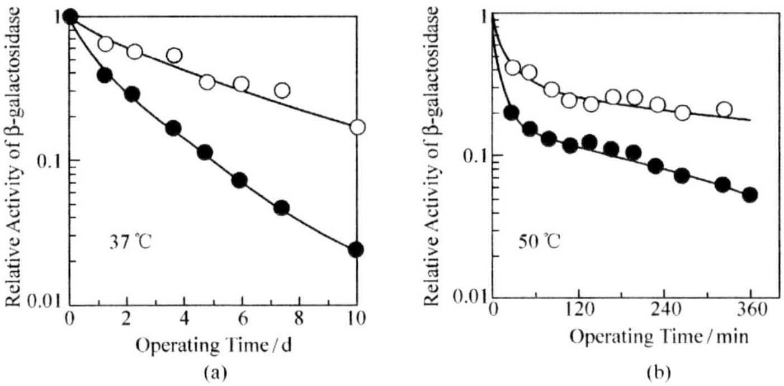

Figure 4.9 Long term stabilities of free β-galactosidase and immobilized derivatives at 37℃ and 50℃, pH = 7. (○): CTP and (●): CPG

enzymes or the dissolving of the surface layer of carrier materials.

The gradual reduction thereafter can be approximated by the following equation analogous to a monomolecular decay model:

$$q = q_0 \exp(-k_d t)$$

where t is the elapsed operating time and q is the relative activity compared to the initial value. k_d is the specific reduction rate. The k_d values of CTP and CPG at 50℃ were obtained from Fig. 4.9 as 0.094 L/h and 0.19 L/h, respectively. The CTP support was twice as effective in stabilizing the enzyme in comparison with CPG.

4.4.3 Immobilization of Alkalophilic Proteinase

Alkalophilic proteinase from **Streptomyces** sp. had an activity of 23 U/mg where one U is the amount of enzyme that hydrolyzes milk casein to release 1 μmol/min of tyrosine at 30℃ and pH 11 in 0.1 M NaOH – Borax buffer solution. The appearance of tyrosine was followed by measurement of the extinction at 275 nm of the reaction mixture.

Figure 4.10 shows changes in the activities of alkalophilic proteinase immobilized on CTP and CPG, both of which have similar pore sizes (~ 40 nm) and surface areas (~ 50 m^2/g). The following are evident from Fig. 4.10: (1) In a solution of pH = 9, which is the optimal pH for the enzyme, the activity of the CPG derivative was rapidly lost. It was reduced to ~ 20% of the initial after 4 days of operation. On the other hand, the activity of the CTP derivative retained more than 20% of the initial even after 16 days of operation. (2) Degradation of the activity for either derivative at pH 7 is much slower than that at pH 9. The excellent stability of CTP derivatives at pH > 7 is considered to originate from better chemical durability of CTP in alkaline solutions. If a different coupling method, which is appropriate for the present CTP, is chosen, the results are expected to improve.

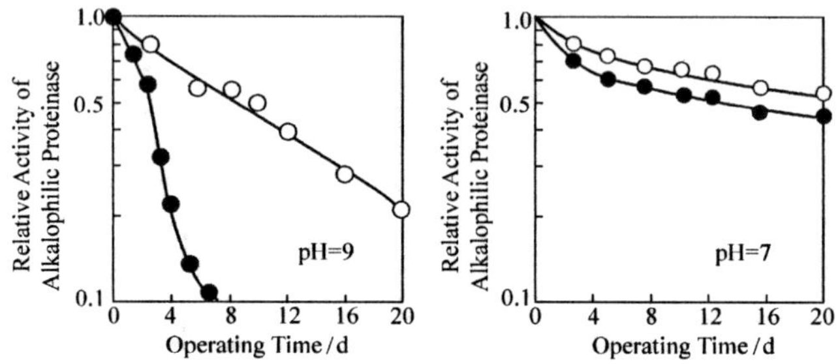

Figure 4.10 Storage stabilities of free alkalophilic proteinase and immobilized derivatives at pH = 9 and 7, 30℃. (○): CTP and (●): CPG

4.4.4 Availability of Porous $CaTi_4(PO_4)_6$ Glass-Ceramics

Figure 4.11 shows a comparison of β-galactosidase and alkalophilic proteinase immobilized onto 100 mg of CTP and CPG, respectively. As shown in Fig. 4.11, the initial activities of CPG derivatives were about four times higher than those of CTP. The CPG derivatives lost their activity rapidly in the initial stages. The reactivities of the CPG derivative became comparable to CTP within 5 days at 37℃. After that, the activity of CTP was always higher than that of CPG (Suzuki et al., 1991).

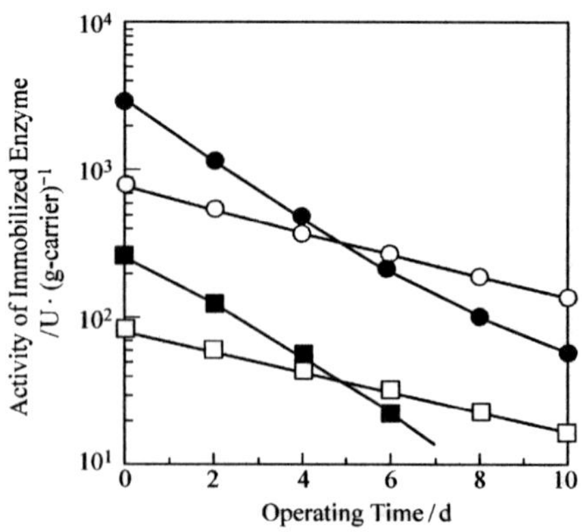

Figure 4.11 Time courses of activities of immobilized β-galactosidase on (○) CTP and (●): CPG; alkalophilic proteinase on (□): CTP and (■): CPG at 37℃, respectively

The immobilizing efficiency and long-term stability of immobilized derivatives are often more desirable economically than quantitative capacity (Chibata, 1989). It is particularly important that CTP derivatives in various pH solutions show high stability.

4.5 Porous Glass-Ceramics with Bacteriostatic Activity

It is known that some silver-or copper-bearing materials have bacteriostatic and fungistatic functions (Goetz, 1943; Simonetti et al., 1992; Russel and Hugo, 1994). Bacteriostatic materials such as zeolites, calcium phosphates, silica gels, and clay minerals which bear Ag^+ ions have recently been commercially used (Korai, 1996). These materials have relatively wide spectra of bacteriostaticity. Against some fungi, copper ions show higher effectiveness in comparison with silver ions (Takayama, 1996). Protons are also effective in bacteriostatic and fungistatic functions (Suzuki, 1984). Since these functions are closely related to the concentrations of these ions around the surface of the specimens, a large surface area is considered one of the important factors in developing materials with high bacteriostatic and fungistatic activity.

4.5.1 Silver-Containing Porous Glass-Ceramics

The authors' group has developed a series of porous glass-ceramics of titanium phosphates. It was found that porous glass-ceramics with a skeleton of $LiTi_2(PO_4)_3$ show excellent cation exchange properties (Hosono et al., 1994) and that Li^+ ions in the glass-ceramics are selectively exchanged with Ag^+ ions even in solutions containing high concentrations of Na^+ ions (Hosono et al., 1994; Hosono and Abe, 1994).

Silver-containing porous glass-ceramics can be prepared by utilizing the ion exchange of Ag^+ ions for Li^+. They show excellent bacteriostatic activity (Hosono and Abe, 1994; Kasuga et al., 1997). The glass-ceramics are bacteriostatic materials which have high thermal resistance and are medically safe since the Ag^+ ions exchanged are chemically stable in water and in the presence of Na^+ ions (Kasuga et al., 1996, 1997).

4.5.1.1 Ion Exchangeability of Glass-Ceramics

There is an increasing demand for cation exchangers in various engineering fields such as biotechnology and environmental engineering to separate or remove a specific ion. Inorganic ion exchangers have several advantages over organic exchangers because of their excellent resistance to heat, chemicals, and radiation. Although a variety of inorganic ion exchangers has been discovered so far (Clearfield and Smith, 1969; Qureshi and Varshney, 1990), there are no materials which can be fabricated via glass, to the authors' knowledge. If an excellent ion exchange material can be synthesized via a glass process, materials can be fabricated in desired shapes and dimensions. Porous ion exchangers are very favorable for application because ion exchange occurs at the surface and near-surface region of materials.

Since the ion conductivities of grain bulk of $LiTi_2(PO_4)_3$ (LTP) ceramics are very high ($\sigma_{300\,K} = 1 \times 10^{-4}\ S \cdot cm^{-1}$) (Aono et al., 1989), there is a likelihood that these materials have excellent ion exchange properties. Porous LTP glass-ceramics show excellent ion exchanging properties, i. e. Li^+ ions in the materials Would exchange with monovalent ions with ionic radius smaller than ~ 130 pm (Na^+, 116 pm; Ag^+, 129 pm; K^+, 152 pm) (Hosono et al., 1993a, 1994).

Figure 4. 12 compares the exchange rates of Li^+ ions with Na^+ ions between porous LTP glass-ceramic and sintered LTP whose surface area is relatively small ($0.3\ m^2/g$). The exchange rate of porous LTP is two orders of magnitude larger than that of the sintered specimen. Approximately 50% of Li ions are exchanged in an hour by the former. On the other hand, the fraction of exchanged Li^+ ions remains only 1% even after 10 h for the latter. The exchange rate of porous LTP glass-ceramics was larger by two orders of magnitude than that of sintered LTP. The value of two in order of magnitude is close to the ratio of specific surface areas between porous and sintered LTPs, which suggests that most of inner

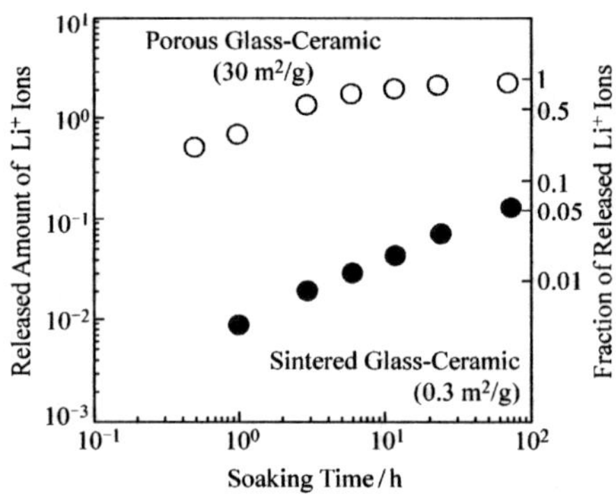

Figure 4. 12 Comparison of exchange rates between porous and sintered LTP glass-ceramics. Solution; 0. 1 *M*$NaNO_3$, temperature 35℃

surfaces of porous LTP are effective for the cation exchange of Li^+ ions. Figure 4. 13 shows the cation exchange rates in porous LTP when specimens are soaked in solutions containing both Na^+ and Ag^+ ions. It is apparent that Li^+ ions in the specimens are selectively exchanged by Ag^+ ions over Na^+ ions. Even in the solution in which the concentration of Ag^+ ions is kept at 1/10 as much as that of Na^+ ions, the fraction of exchanged Ag^+ ions is about five times as large as that of Na^+ ions. The exchange rate of Na^+ ions with Li^+ ions in the LTP is larger than that of Ag^+ ions. The strong affinity of Ag^+ over Na^+ in mixed ion states may be attributed to the instability of Na^+ ions occupying Li^+ sites over Ag^+ ions and is considered to originate from the difference in ionic radius ($R_{Ag^+} - R_{Li^+} > R_{Na^+} - R_{Li^+}$).

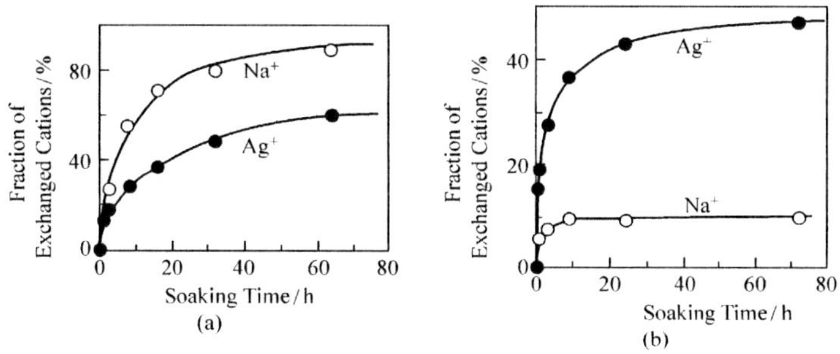

Figure 4.13 Changes in fraction of Na^+ and Ag^+ ions exchanged with Li^+ ions in porous LTP glass-ceramics. (a) in aqueous solutions of 0.1 $M$$NaNO_3$ or 0.1 $M$$AgNO_3$, (b) in aqueous solution containing both $[Na^+] = 0.1\ M$ and $[Ag^+] = 0.01\ M$

4.5.1.2 Effects of Chemical Composition on Pore Properties

Figure 4.14 shows the variation in glass transition temperature (Tg) and crystallization temperature (Tc), which are estimated from DTA, with the

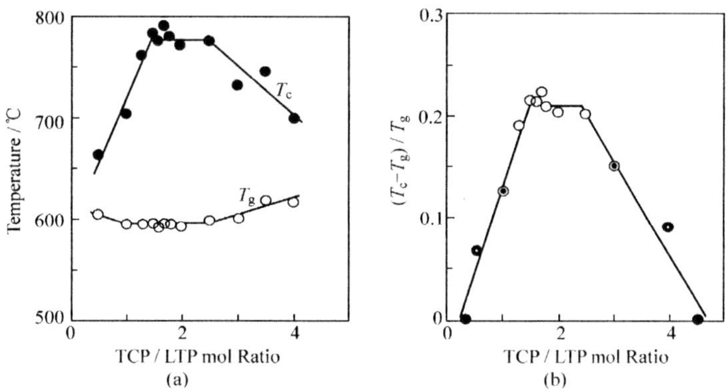

Figure 4.14 Effects of chemical composition on glass transition temperature (Tg) and crystallization temperature (Tc). (right); (○): clear glass, (⊙): glass + crystal (trace), (**○**): glass + crystal, and (●): crystal

nominal composition of the mother glasses (Hosono and Abe, 1998). The parameter defined as (Tc − Tg) / Tg, which is an indicator of the thermal stability of a glass, has a maximum around the ratio of 1.5 − 2.0. This range agrees well with that for a clear glass. The pore size distribution curve was nearly log normal in shape for each specimen. The porosity was 50 vol% − 60 vol%, irrespective of the composition of the mother glasses. When the concentration ratio of $Ca_3(PO_4)_2$(TCP) to $LiTi_2(PO_4)_3$(LTP) changes from 1.5 to 3.5, the specific surface area increases from 15 − 25 m^2, and the median pore diameter decreases from 300 to 100 nm, as shown in Fig. 4.15.

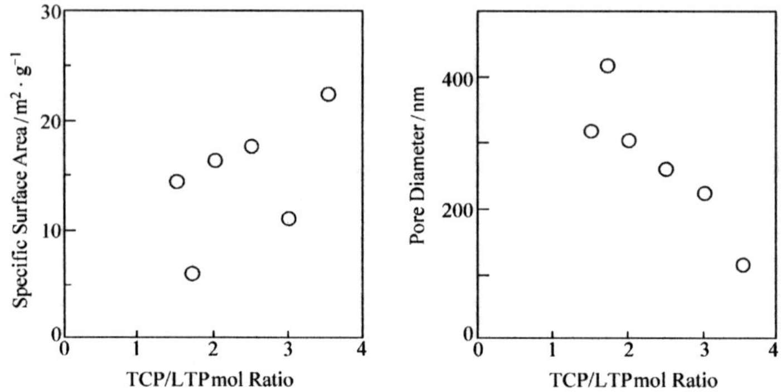

Figure 4.15 Variation in specific surface area and median pore diameter of porous LTP glass-ceramics with the composition of the mother glasses. TCP: $Ca_3(PO_4)_2$ and LTP: $LiTi_2(PO_4)_3$

4.5.1.3 Bacteriostatic Activity

High selectivity to Ag^+ ions resulting from LTP crystals and a high exchange rate due to large surface areas will be very advantageous for these applications. Specimens can be obtained from a mother glass without accompanying serious cracking, shrinkage, and deformation. The outstanding shaping of the glass-ceramic process will render the present materials flexible to meet various specifications.

The pore diameter of porous LTP glass-ceramics was ~ 60 nm, which is so small that diffusion of ions in pores would influence the ion exchange rate. A larger pore size and superior crystallographic modification are needed to enhance the diffusion coefficient of cations and thereby increase the ion exchange rate.

When the crystal forming skeleton of porous glass-ceramics is changed from $LiTi_2(PO_4)_3$ to $Li_{1.4}Ti_{1.6}Al_{0.4}(PO_4)_3$ [denoted by LATP, electrical conductivities in this solid solution are larger by an order of magnitude than those in an LTP crystal (Hosono and Abe, 1991a)], the pore size is increased by a factor of three (Hosono and Abe, 1992b), and the cation exchange rate is significantly enhanced. Figure 4.16 shows scanning electron micrographs (SEM photos) of porous LTP and LATP glass-ceramics derived from glasses with compositions of TCP/LTP and TCP/LATP ratio = 1.1, respectively. The pore size in LATP is considerably larger than that in LTP.

A 0.5 g sample of the porous glass-ceramic grains was soaked in 50 mL of aqueous solution with 0.1 *M*$AgNO_3$ at room temperature for 3 days. Approximately 1.5 – 2 mequiv/g of Ag^+ ions were introduced into the porous glass-ceramics during this treatment (Hosono et al., 1994). As a result, porous glass-ceramics with a surface phase of a majority of $AgTi_2(PO_4)_3$ crystal and an

(a) (b)

Figure 4.16 SEM photos of porous (a) LTP and (b) LATP glass-ceramics derived from the glasses with compositions of TCP/LTP and TCP/LATP ratio = 1.1, respectively

interior phase of $LiTi_2(PO_4)_3$:Al crystal were prepared by ion exchange of Ag^+ ions for Li^+ ions (The resultant material is denoted by porous glass-ceramic Ag-LATP). The Ag-LATP glass-ceramic exhibits excellent antibacterial activities to various bacteria.

Antibacterial activities of the specimens were measured for **Escherichia coli** (IFO 3208), **Staphylococcus aureus** (IFO 13276), **Pseudomonas aeruginosa** (IFO 17018), and **Streptococcus mutans** (IFO 13955). Bacterial cell suspensions were added to phosphate buffer (50 mM, pH 6.0) containing 0.005 wt% of the Ag-LATP glass-ceramic and kept at 37℃, and the number of living bacterial cells was counted after 6 h, 24 h, and 48 h.

Table 4.2 summarizes the results of the examination of the antibacterial activities. The number of living bacterial cells was reduced to the detection limit in 6 h for all the kinds of bacteria examined here.

Table 4.2 Changes in viable counts of bacteria with time

Bacteria	0 h*	6 h	24 h	48 h
Escherichia coli (IFO 3208)	6.3×10^5	< 1	< 1	< 1
Staphylococcus aureus (IFO 13276)	9.2×10^5	< 1	< 1	< 1
Pseudomonas aeruginosa (IFO 17018)	8.4×10^5	< 1	< 1	< 1
Streptococcus mutans (IFO 13955)	1.2×10^5	< 1	< 1	< 1

* The numbers of living cells at "0 h" were counted immediately after bacterial cell suspensions were added to phosphate buffer solution containing the glass-ceramics

The amount of Ag^+ ions released from the as-prepared Ag-LATP into DW was very small (0.4 – 0.6 μequiv/g), while the amount released into phosphate buffer solution containing NaCl was larger than that into DW. The release in phosphate buffer solution, however, can be drastically suppressed by heating the porous LATP glass-ceramics (Kasuga et al., 1996, 1997). Figure 4.17 gives plots of the amount of Ag^+ ions released into distilled water and phosphate buffer solution from the porous Ag-LATP glass-ceramic and from that heated at 900℃. Even after soaking for 0.25 h, the Ag-LATP glass-ceramic released Ag^+ ions of ~30 μequiv/g, and further increase in the amount of the released ions was not seen even after prolonging the soaking time. The amount of the Ag^+ ions released into phosphate buffer solution was measured to be 1 – 2 μequiv/g and was an order of magnitude smaller than that from as-prepared porous Ag-LATP glass-ceramic without heating. The amount of ions released into DW also was very small. XRD analysis showed that the skeleton of $(Ag,Li)Ti_2(PO_4)_3$: Al crystal in the glass-ceramic is not changed by heating at 900℃.

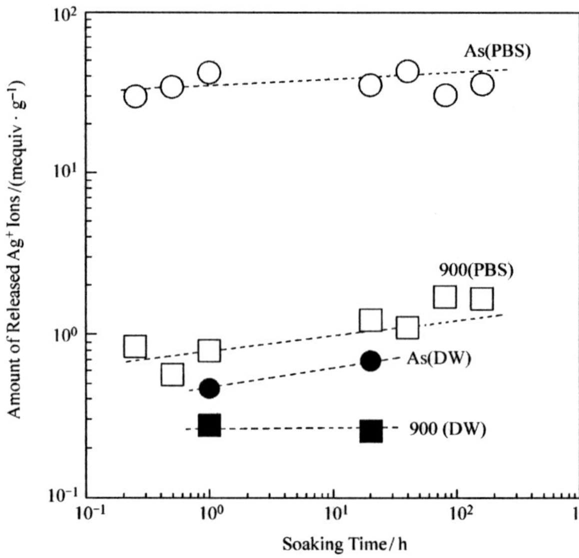

Figure 4.17 Amounts of Ag^+ ions released into DW or phosphate buffer solution (PBS) containing Na^+ ions from porous glass-ceramics Ag-LATP before and after heating as a function of soaking time. As(PBS): the amount released into PBS from porous Ag-LATP glass-ceramic; As(DW): the amount released into DW from the Ag-LATP glass-ceramic; 900(PBS): the amount released into PBS from the Ag-LATP glass-ceramic heated at 900℃, and 900(DW): the amount released into DW from the Ag-LATP glass-ceramic heated at 900℃

The porous Ag-LATP glass-ceramic showed specific surface areas of 9 – 10 m^2/g, and, when it was heated at 900℃, the area was decreased to ~ 5 m^2/g. Figure 4.18 shows SEM photos of the Ag-LATP glass-ceramic heated at

900℃. The grains consisting of $(Ag,Li)Ti_2(PO_4)_3$:Al crystals are roundish, and some portions of them are sintered. The decrease in the surface area is due to the sintering behavior of the glass-ceramic. Although the surface area of the glass-ceramic Ag-LATP heated at 900℃ was decreased by ~1/2 as much as that of the as-prepared Ag-LATP, the release amount of Ag^+ ions into PBS was decreased by ~1/30. It is difficult to suggest that the decrease of the surface area is closely related to the release amount of Ag^+ ions. The mechanism for the reduction of the amount is not clear yet.

Figure 4.18 SEM photos of porous Ag-LATP glass-ceramic heated at 900℃

The effectiveness of porous Ag-LATP glass-ceramics heated at 900℃ against Escherichia coli (IFO 3301) was examined. The number of living bacterial cells in the suspensions containing the heated sample was reduced to <1 cells: its excellent bacteriostatic activity was confirmed. This fact also means that the Ag-LATP glass-ceramic does not lead to deterioration of activity even after heating at high temperature (e. g., 900℃). As the amount of Ag^+ ions released into phosphate buffer solution from the porous Ag-LATP glass-ceramic heated at 900℃ is very small (1 – 2 μequiv/g), the material is expected to be safe to our bodies.

As anticipated, partially Ag^+-exchanged porous LATP glass-ceramics exhibited excellent antibacterial activities. Since antibacterial activity increases with concentration of Ag^+ ions on surfaces in contacting with the environment, the realization of higher Ag^+ concentrations at surfaces and near surface regions is very favorable with respect to cost performance. In addition, since Na^+ ions are one of the most abundant ions in our environment, the materials are in most cases used in the presence of Na^+ ions. Thus, Ag^+ ions in the material need to be stable in such an environment. The present porous LATP glass-ceramics meet the above requirements. Silver ions can be easily inserted into surfaces and near surface regions of the materials by partial ion exchange of Li^+ with Ag^+ and concentrations of Ag^+ ions can be raised to ~ 60 wt% [corresponding to $AgTi_2(PO_4)_3$]. Exchanged Ag^+ ions are chemically stable in water and even in the presence of Na^+ ions. Since the present porous materials are prepared by the glass-ceramic process, materials can be obtained in desired shapes and dimensions. Outstanding dimensional precision will render the present materials flexible to meet various specifications.

4.5.2 Porous Glass-Ceramics with a Skeleton of Copper Titanium Phosphate

The authors prepared a porous glass-ceramic with a skeleton of Nasicon-type cupric titanium phosphate $CuTi_4(PO_4)_6$ (denoted by Cu^{II}TP) (Oudet et al., 1989) showing high catalytic activity in the dehydrogenation and dehydration of 2-propanol (Yamamoto and Abe, 1998). Cu^{2+} ions and protons introduced during the acid-leaching process are suggested to be located in the sites of the conduction channels.

As mentioned above, copper ions show higher effectiveness in comparison with silver ions against some fungi (Takayama, 1996). Protons are also effective in bacteriostatic and fungistatic functions (Suzuki et al., 1984). Porous Cu^{II} TP glass-ceramic may meet the above requirements. There inevitably exist Na^+ ions in an environment in which bacteria or fungi grow. If protons or copper ions can be released in response to the presence of Na^+ ions, glass-ceramic Cu^{II} TP is expected to be applicable as a novel bacteriostatic and fungistatic material with the capacity to adapt to the environment.

4.5.2.1 Preparation of Porous $CuTi_4(PO_4)_6$ Glass-Ceramics

A nominal composition of $50CuO \cdot 20TiO_2 \cdot 30P_2O_5$ (mol%), which was melted in an alumina crucible in air, was quenched in a nitrogen atmosphere after the glass was heated at 520 – 590℃. Figure 4.19 shows the XRD pattern of the resulting dense glass-ceramic. $Cu(I)Ti_2(PO_4)_3$, $Cu_3(PO_4)_2$, and CuO crystals were precipitated in the glass (Yamamoto et al., 1997).

The dense glass-ceramics were converted into porous bodies with the original shape after leaching with 0.5 *N* H_2SO_4 or 0.5 *N* HCl solution at room temperature. Figure 4.20 shows the SEM photos of the fracture face and the XRD patterns of the samples after leaching with the acid solutions. The crystals of $Cu_3(PO_4)_2$ and CuO were dissolved completely. The porosity, pore size, and density of the porous samples obtained are summarized in Table 4.3. The specific surface areas are about 40 – 50 times larger than those prepared by a conventional sintering method (Serghini et al., 1992). XRD analysis showed that the sample leached with H_2SO_4 consists of $Cu(I)Ti_2(PO_4)_3$ (denoted by Cu^ITP) crystal (Mbandza et al., 1985) and that the sample leached with HCl

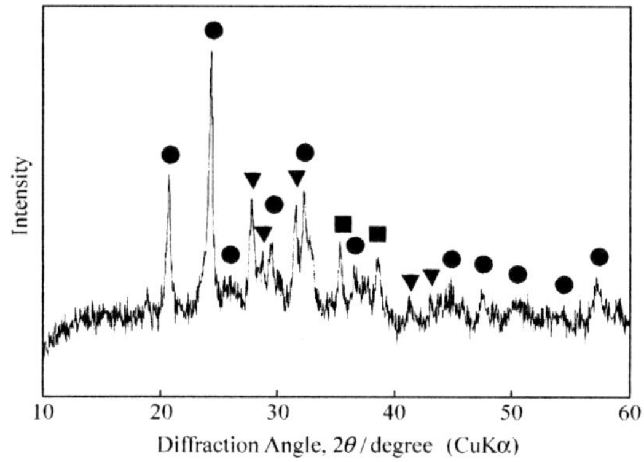

Figure 4.19 Powder XRD pattern of Cu – Ti – P – O glass-ceramic prepared by heating at 520℃, followed by heating at 550℃ and 590℃. (●): Cu(Ⅰ)$Ti_2(PO_4)_3$, (▼): $Cu_3(PO_4)_2$ and (■): CuO

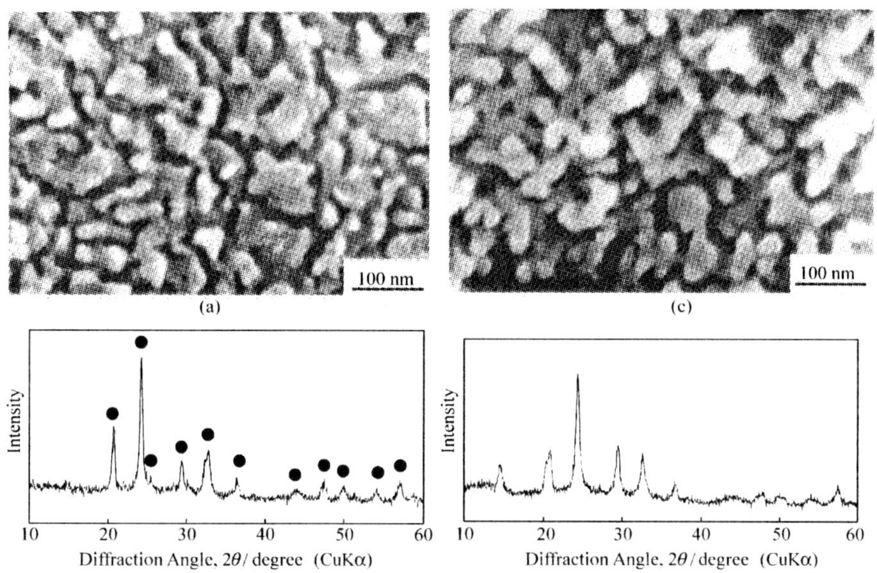

Figure 4.20 SEM photos of fracture faces and powder XRD patterns of the glass-ceramics after leaching with 0.5 *N* H_2SO_4 and 0.5 *N* HCl. (a), (b): sample leached with H_2SO_4 and (c), (d) sample leached with HCl. (●) in (b): Cu(I)$Ti_2(PO_4)_3$

consists of Cu^{II} TP crystal, judging from the pattern calculated from high temperature neutron diffraction (Olazcuaga et al., 1994).

Their colors changed to brownish or bluish green by leaching with H_2SO_4 and HCl, respectively. The ESR spectrum in Fig. 4.21(a) indicates that the glass-ceramic after HCl leaching contains a large amount of copper(Ⅱ) ions. The ESR parameters of copper(Ⅱ) ions, $g_{//} = 2.33$ and $g_{\perp} = 2.06$, indicate

Table 4.3 Physical properties of porous Cu – Ti – P – O glass-ceramics prepared by leaching with H_2SO_4 or HCl

	Porous glass-ceramic obtained by H_2SO_4 leaching	Porous glass-ceramic obtained by HCl leaching
BET surface area $/m^2 \cdot g^{-1}$	46.0	70.0
Average pore radius /nm	9.3	7.2
Pore volume $/cm^3 \cdot g^{-1}$	0.2	0.3
Porosity (vol%)	38.0	50.0
Bulk density $/g \cdot cm^{-3}$	2.0	1.5
True density $/g \cdot cm^{-3}$	3.2	3.1

the possibility of the presence of a hyperfine structure belonging to copper(II) ions. On the other hand, as shown in Fig. 4.21 (b), the glass-ceramic after H_2SO_4 leaching contains a trace amount of copper(II) ion.

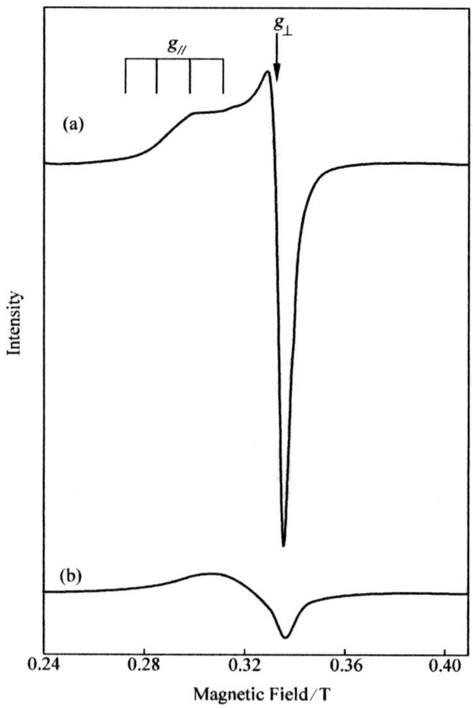

Figure 4.21 ESR spectra at room temperature in (a) the porous glass-ceramic obtained by leaching with 0.5 *N* HCl and (b) the porous glass-ceramic obtained by leaching with 0.5 *N* H_2SO_4

Quantitative analysis by inductively coupled plasma atomic emission spectroscopy (ICP – AES) showed that the content of copper ions in the porous glass-ceramic prepared by HCl leaching is less than that in the Cu^{II} TP crystal. Some copper ions may be exchanged for H^+ ions in the HCl solution, owing to the channeling structure of the Nasicontype (Warner et al., 1992). The porous glass-ceramic prepared by HCl leaching is suggested to include copper(II) ions and protons in the conduction channel (denoted by porous Cu^{II} TP glass-ceramic).

In general, when the glass is melted at a high temperature (e. g., > 1000℃), the copper(I) state is thermodynamically stable and incorporated in glass during quenching (Kamiya et al., 1986; Lee et al., 1993). Therefore, Cu^{I}TP can be directly crystallized from the parent glass. The $Cu_3(PO_4)_2$ crystal in the mother glass-ceramic is preferentially dissolved in acid solution. The Cu^{I}TP crystal was retained in H_2SO_4 solution. On the other hand, when HCl was used as the leaching solution, the Cu^{I}TP crystal was converted into Cu^{II} TP. Since the equilibrium concentration of copper(I) ions is very small in acid ic aqueous solutions ($< 10^{-2}$ M), the copper(I) ions form a complex that dissolves in the solution (Cotton et al., 1976; Cotton and Wilkinson, 1981). Owing to this complex-forming ability of Cl^- ions, the copper(I) ions in Cu^{I}TP crystal dissolve successfully in HCl solution. The skeleton of the Nasicon-type structure is appropriate for the dissolution of copper(I) ions (Olazcuaga et al., 1994), since the three-dimensional rigid framework is not destroyed during acid leaching and gives channels for the motion of copper(I) ions (Warner et al., 1992). On the other hand, no SO_4^{2-} ions act as a ligand for copper(I) ions. Owing to the channeling structure, the H^+ ions are introduced into the vacant sites of the copper(I) ions and compensate for the electric charge valence. This preparation method is a novel process of the oxidation of copper (I) ions in the Nasicon-type titanium phosphate crystal into copper(II) ions without heating under an oxygen-containing atmosphere.

4.5.2.2 Control of Ion Release

The release of excessive ions may shorten the life of the materials and contaminate the environment. The released amount of cupric ions was found to be restricted drastically by doping the Nasicon-type crystal Cu^{II}TP with Ca^{2+} ions.

Glasses with a nominal mol% composition of $(50-x)CuO \cdot xCaO \cdot 20TiO_2 \cdot 30P_2O_5$ ($x = 0-2.5$) were heated at 510 – 590℃, resulting in dense glass-ceramics, and subsequently HCl leaching was carried out to obtain porous Cu^{II} TP glass-ceramics. Ca^{2+} ions are located in the sites of the conduction channels of the Nasicon-type crystal. As described in Sect. 4.5.2.1, porous Cu^{II} TP glass-ceramics include protons, that is, Cu^{2+}, H^+, and Ca^{2+} ions are located in the conduction channels of the porous Cu^{II} TP glass-ceramics. Their specific surface areas and average pore radii are 50 – 60 m^2/g and ~ 20 nm, respectively, and the area and radius are not influenced by the CaO content in the experimental range.

Figure 4. 22a shows the amounts of copper and phosphorus ions released into 0. 1 M NaCl aq. from the porous glass-ceramics and the pH values of the solution. The amount of copper ions released increased with increasing soaking time, as did that of phosphorus ions. The amounts were reduced dramatically by doping Ca^{2+} ions into the Nasicon-type crystal. The amount of copper ions released from the porous glass-ceramic derived from the mother glass with a composition of $x = 1.5$ in $(50 - x)CuO \cdot xCaO \cdot 20TiO_2 \cdot 30P_2O_5$, was smaller by ~1/3, and that of $x = 2.5$ was an order of magnitude smaller than that of $x = 0$. The released behavior of phosphorus ions was also similar to that of copper ions. The concentrations of titanium and calcium ions dissolved in the solution were estimated to be < 10 μequiv/g. On the other hand, when the porous glass-ceramics were soaked in DW, almost no ions were dissolved. As shown in Fig. 4. 22b, the pH of the NaCl solution decreased quickly after the glass-ceramics were introduced. A large amount of H^+ ions was dissolved from the glass-ceramics. The dissolution of ions from the porous glass-ceramics occurs in the presence of Na^+ ions.

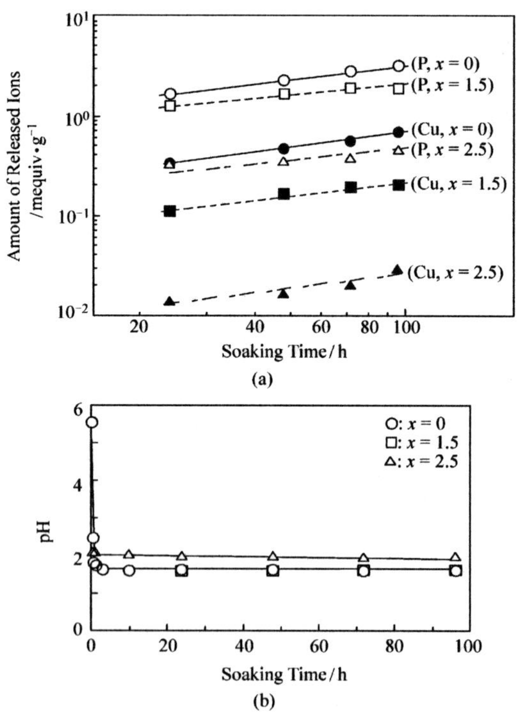

Figure 4. 22 (a) Amounts of copper and phosphorus ions released into 0. 1 M NaCl aq. (50 mL) from the porous glass-ceramics and (b) pH values of the solution. Cu and P in parentheses indicate the released Cu^{2+} and P^{5+} ions, respectively. $x = 0$, 1. 5, and 2. 5 correspond to the glass-ceramics derived from the mother glass with a composition of $(50 - x)CuO \cdot xCaO \cdot 20TiO_2 \cdot 30P_2O_5$

When the glass-ceramics are soaked in Na^+-ion-containing solution, the Na^+ ions are exchanged for H^+ ions which are considered to be located in some conduction channels of the Nasicon-type $Cu^{II}TP$ crystal. The H^+ ions with high mobility are dissolved successively from the glass-ceramics by ionic exchange. On the other hand, dissolved H^+ ions may chemically erode the surface of the materials, whereby copper and phosphorus ions are gradually released. Since the solubility of titanium ions is very low in the solution, they may be condensed to form related phases or be precipitated as phases in the solution. The mobility of Ca^{2+} ions with a large ionic radius is very low (Mentre et al., 1994). When Ca^{2+} ions are included in the $Cu^{II}TP$ crystal, the mobility of H^+ ions is thought to be affected. Then, the rate of exchange of H^+ ions for Na^+ ions would be reduced. As a result, the dissolution of copper ions is suggested to be restricted.

The efficiency of the porous glass-ceramics against Escherichia coli (IFO 3301) was examined. The number of living bacterial cells in the suspension containing the porous glass-ceramic derived from the glass with $x = 0$ and 1.5 was determined to be <1, and that for $x = 2.5$ was reduced to 0.1×10^5. Their excellent bacteriostatic activities were confirmed. The activities are the result of the dissolution of H^+ and Cu^{2+} ions. Porous glass-ceramics $Cu^{II}TP$ are smart materials which can automatically control the release of H^+ and Cu^{2+} ions in response to Na^+ ions in the environment (Kasuga et al., 1999a, 1999b).

4.6 Porous Glass-Ceramics with an Integrated Skeleton

Recently, to meet the demands of environmental chemical technologies, a great deal of attention is being paid to bacteriostatic materials with various properties such as adsorption or decomposition of odorous products, water repellence, or good compatibility with plastic material. It is expected that such materials can be used for various applications such as wallpapers, tiles, and plaster. The above mentioned bacteriostatic glass-ceramics have low or almost no activities such as adsorption, decomposition, or water repellence.

As described in Section 4.3, porous glass-ceramics with a skeleton of γ-$Ti(HPO_4)_2 \cdot 2H_2O$ (THP) crystal which have a two-dimensional layered structure (Clearfield, et al., 1973; Arberti, 1978; Clearfield and Roberts, 1988) can be prepared by acid treatment of the resulting glass-ceramics composed of $LiTi_2(PO_4)_3$ (LTP) and β-$Ca_3(PO_4)_2$ (TCP) (Hosono and Abe, 1992a). The porous glass-ceramics have large surface areas ($\sim 120\ m^2/g$) and can be modified chemically by intercalation of organic polar molecules. Although the proton in THP crystal is easily exchanged for Ag^+ or Na^+ ions, these immobilized ions are released often when the environment such as pH is changed.

The present section describes a new type of multifunctional microporous material (Kasuga et al., 1998) with an integrated skeleton of $AgTi_2(PO_4)_3$ and $Ti(HPO_4)_2 \cdot 2H_2O$ (THP) crystals with multifunctionality such as bacteriostatic activity, adsorption of ammonia gas, and intercalation of organic polar molecules.

4.6.1 Preparation Procedure

A plate of porous $LiTi_2(PO_4)_3$:Al (LATP) glass-ceramic was soaked in 50 mL of aqueous solution with 0.1 *M* $AgNO_3$ at room temperature for 72 h. As a result, porous glass-ceramics with a surface phase consisting predominantly of $AgTi_2(PO_4)_3$ and an interior phase of LATP in the crystalline skeleton were prepared by ion exchange of Ag^+ ions for Li^+. The ion-exchanged products were heated at 900℃ for 1 h since the amount of Ag^+ ions released from them the into aqueous solutions was very small (Kasuga et al., 1997). This resultant material is denoted by porous Ag-LATP glass-ceramic, hereafter. The specific surface area and median pore diameter of the Ag-LATP was estimated to be ~ 5 m^2/g and 0.2 μm by the BET method, respectively.

Next, the porous Ag-LATP glass-ceramics were soaked in 1 *N* HNO_3 at the boiling temperature (~ 100℃) under stirring and refluxing to form THP crystal. During boiling, Li^+ ions with high mobility in a Nasicon-type phase in the porous Ag-LATP glass-ceramic are easily exchanged for protons or oxonium ions. Subsequently, the resulting products are converted into a THP phase by a chemical reaction such as hydration (Hosono and Abe, 1992b) although the formation mechanism is not clear yet. The mobility of Ag^+ ions in the Nasicon-type $(Ag, Li)Ti_2(PO_4)_3$ crystal is suggested to be relatively small (Hosono et al., 1994; Kasuga et al., 1998). When Ag^+ ions are introduced into a conduction site in a Nasicon-type phase, the mobility of Li^+ ions in the phase is reduced. As a result, the phase is converted slowly into a THP phase and the $AgTi_2(PO_4)_3$ phase is left alone. Therefore, by boiling the porous Ag-LATP glass-ceramic, a material consisting mainly of $AgTi_2(PO_4)_3$ and THP phases is prepared.

Although the BET surface area of the Ag-LTP glass-ceramic before treatment was ~ 5 m^2/g, it was increased dramatically by increasing the boiling time (e.g., ~ 30 m^2/g after treatment for 12 h). The increase in area originates from the formation of the THP phase. No significant deformation of the bulk samples occurred after boiling.

Figure 4.23 shows SEM photos of a sample obtained by boiling the Ag-LATP glass-ceramic. As shown in Fig. 4.18 in Sect. 4.5.1.3, the grains consisting of $(Ag,Li)Ti_2(PO_4)_3$:Al crystals are roundish and are connected to each other, resulting in the formation of pores 0.2 – 0.5 μm in size. It is seen in Fig. 4.23 that the porous material with a skeleton consisting of angular grains is formed after boiling: no serious fracture of the skeleton is seen, and the shape and size of the observed macropores are close to those before treatment. With the

Figure 4.23 SEM photos of the porous Ag-LATP glass-ceramic treated with 1 N HNO_3 at 100℃ for 12 h.

boiling, the LATP phase which exists in the interior of the skeleton of the Ag-LATP glass-ceramic is converted into a THP phase, while the surface layer phase $AgTi_2(PO_4)_3$ is not changed. Figure 4.24a – c shows pore size distribution curves of a porous Ag-LATP glass-ceramic, the porous THP glass-ceramic (Hosono and Abe, 1992b) prepared by the mother (dense) glass-ceramic, and

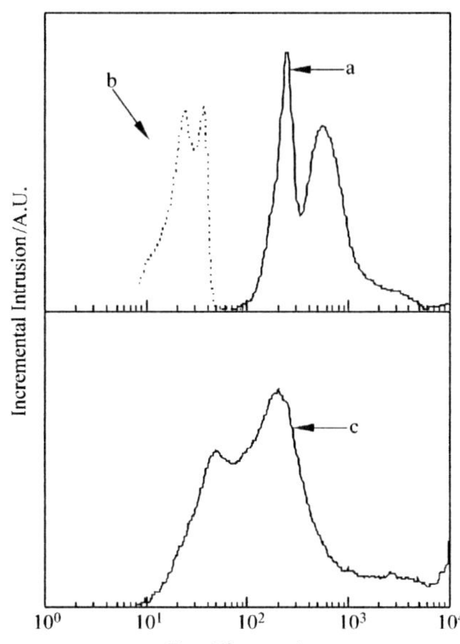

Figure 4.24 Pore size distribution of (a) porous glass-ceramic Ag-LATP, (b) porous THP glass-ceramic, and (c) the materials obtained by boiling porous Ag-LATP glass-ceramic for 12 h

the material prepared by boiling the Ag-LATP glass-ceramic, respectively. The distribution of the Ag-LATP glass-ceramic is bimodal, and two peaks are located at ~ 200 nm and ~ 600 nm (Fig. 4.24a). The THP glass-ceramic is a material with mesopores 10 – 50 nm in size (Fig. 4.24b). As shown in Fig. 4.24c, the curve of the material prepared by boiling the porous Ag-LATP glass-ceramic shows a wide pore size distribution ranging from 10 nm to 1 μm and has two main peaks located at 200 nm and 40 nm. A superimposition of traces [(a) and (b) in Fig. 4.24] of porous glass-ceramics Ag-LATP and THP is very close to curve (c). Since the THP phase was formed by boiling the Ag-LATP glass-ceramic, pores ranging from 10 nm to 100 nm appeared newly in addition to those 100 nm – 1 μm. The material is suggested to be composed of an integrated skeleton of crystalline γ-$Ti(HPO_4)_2 \cdot 2H_2O$ with mesopores and $AgTi_2(PO_4)_3$ phases.

4.6.2 Functions of Glass-Ceramics

4.6.2.1 Bacteriostatic Activities

The effectiveness of porous material with an integrated skeleton against E. coli (IFO 3301) was examined using the method shown in Sect. 4.5.1.3 (Kasuga et al., 1998). The numbers of living bacterial cells counted in the suspension containing the powdery glass-ceramic after incubation for 20 h were < 1. The excellent bacteriostatic activity of the porous material with the integrated skeleton was confirmed as well as that of the porous glass-ceramic Ag-LATP. The porous THP glass-ceramic did not show bacteriostatic effectiveness. The activity of the material is attributed to the $AgTi_2(PO_4)_3$ phase in the integrated skeleton. The amount of Ag^+ ions released into phosphate buffer solution by soaking the sample for 20 h at 37℃ was determined to be ~ 8 μequiv/g. Since the amount is very small, the material is expected to have long lifetime of bacteriostatic activity and to be safe to our bodies. A 50% lethal dose (LD_{50}) of $AgNO_3$ is reported to be 50 mg/kg (Sweet, 1985).

4.6.2.2 Adsorption of Ammonia Gas

The adsorption activity of ammonia as one of the typical odorous gases was evaluated. A 5 mg sample of grains sieved from 104 to 208 μm was set around the center in a closed silica glass tube with a volume of 723 mL. A difference between the initial amount (257 ppm) of ammonia gas and the measured, residual gas was defined as the gas adsorbed on the sample. (Kasuga et al., 1998)

The adsorption activity of ammonia gas on the porous materials prepared by boiling are shown in Fig. 4.25. The adsorption to the Ag-LTP glass-ceramic was as effective as that of activated charcoal, while that of the materials prepared by boiling was improved dramatically. Ammonia gas at a concentration of

~200 ppm can be adsorbed using 5 mg of the materials within 5 min. The activity is based on the formation of P – OH_4N by reaction between NH_3 and the P – OH group located at the interlayers (Clearfield and Hunter, 1976; Tsuhako et al., 1988) in THP phase.

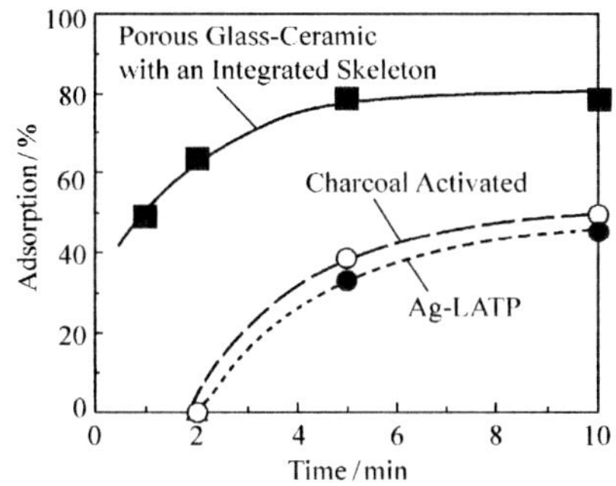

Figure 4.25 Ammonia adsorption of activated charcoal, porous Ag-LATP glass-ceramic, and porous glass-ceramic with the integrated skeleton obtained by treatment with 1 *N* HNO_3 at 100℃ for 12 h

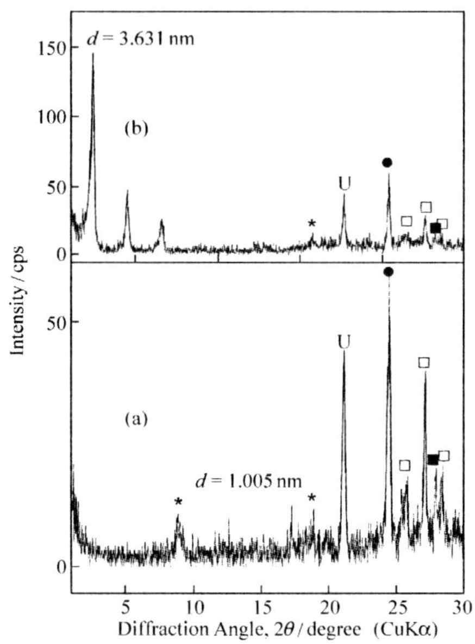

Figure 4.26 XRD patterns of the porous material with the integrated skeleton prepared by boiling for 12 h (a) before and (b) after soaking in a methanol solution of dodecylamine. (●); $AgTi_2(PO_4)_3$, (*); $Ti(HPO_4)_2 \cdot 2H_2O$, (■); $Ti(OH)PO_4$, (□); $(TiO)_2P_2O_7$, and (U); unknown phase

4.6.2.3 Intercalation of Polar Molecules

The porous glass-ceramic with the intergrated skeleton was soaked in methanol solutions of monoalkylamines at room temperature. Figure 4.26 shows the change in XRD patterns of the porous glass-ceramic with the integrated skeleton and of that after soaking in a methanol solution of dodecylamine. The main peak of the THP phase in the material shifted from $d = 1.005$ nm to 3.631 nm. This shift corresponds to an increase in the interlayer spacing as a result of the intercalation of dodecylamine into the interlayers of the THP phase. P – OH groups in the THP phase are available to link amines (Alberti and Constantino, 1978; Moya, 1979) Monoalkylamines are intercalated in a state of the bilayer structure at an inclined angle of ~ 58° into the interlayers in the THP phase in the material with the integrated skeleton (Kasuga et al., 1998). Further evidence shows that the integrated glass-ceramic has hydrophilicity (bulk density; ~1.5 g/mL) because it sinks when put into water. Contrary to this, the amine-intercalated samples show hydrophobicity by floating on water. The amine-intercalated samples showed excellent water-repellent properties.

4.7 Concluding Remarks

This chapter introduced some novel porous composite materials developed by utilizing a glass-ceramic technique to be applied to the biotechnology field. The porous glass-ceramics have higher performance in comparison with conventional materials, because they have the combined advantage of porous ceramics (large surface area), the glass-ceramic process (process merits), and crystal skeleton (functionality).

A variety of applications, including matrix materials for studying the mesoscopic phenomena (Orazio et al., 1989; Borisov et al., 1994) of this novel class of porous glass-ceramics, are expected utilizing these unique features. The authors examined applications utilizing the characteristics of each porous glass-ceramic. Table 4.4 summarizes the characteristics of porous glass-ceramics prepared, so far, together with their applications (Hosono and Abe, 1995).

Porous glass-ceramics analogous to the present examples may be able to be fabricated in many other systems. The primary point in the design of novel porous-glass-ceramics is to choose an appropriate combination of a functional skeleton crystal and its countercrystal, $RTi_2(PO_4)_3$ and $Ca_3(PO_4)_2$ in the present case. Desired chemical compositions of this combination need to meet the following two requirements: (a) stable glasses must be obtained and (b) resulting glasses are able to be converted into glass-ceramics via spinodal-type phase separation. The sol-gel process utilizing a mixture of organic polymers and polymerizing inorganics (Nakanishi and Soga, 1991) may be promising for this

Table 4.4 Porous glass-ceramics prepared and their applications

Type	Skeleton	Median pore diameter/nm	Specific surface area/m^2/g	Porosity (vol%)	Properties	Applications
I						
Nasicon-type (3-D structure)	$CaTi_4(PO_4)_6$[1]	40 – 60 (adjustable)[2]	50 – 90	50	Alkali durability	Immobilized enzyme[3]
					Weak affinity to water, low thermal expansion	Humidity sensor[4]
	$NaTi_2(PO_4)_3$[5]	70	45	50	Zero thermal expansion	
	$Li_{1-x}Ti_{2-x}Al_x(PO_4)_3$[6]	200	50	55	fast Li^+ conductor[7] Ag^+-selective ion exchanger[7,8,9]	Bacteriostatic/fungistatic materials[9,10]
	$CuTi_2(PO_4)_3$[11]	20	45	40	Catalytic activity[11,12]	Oxidation catalyst of propene[11,12]
	$CuTi_4(PO_4)_6$[13]	15	70	50	Catalytic activity[13]	Decomposition of 2-propanol[13]; bacteriostatic/fungistatic materials[14]
II						
Zirconium phosphate-type (2-D structure)	$Ti(HPO_4)_2 \cdot 2H_2O$[15]	15 – 20	120	45	Ion exchanger	Cationic exchanger
					Affinity to proteins; intercalatable with polar organic molecules[15]	Protein purification
III						
3-D/2-D Integrated	$LiTi_2(PO_4)_3/Ti(HPO_4)_2 \cdot 2H_2O$[16,17]				Combined properties of type I and II	Intelligent bioreactors
IV						
	$TiO_2 + \alpha$-SiO_2[18]	1200	40	65	Large pore diameter stable at high temperature[19]	O_2 sensor of car exhaust gases[19]

Ref. [1] Hosono et al., 1989; [2] Hosono et al., 1992; [3] Suzuki et al., 1991; [4] Hosono and Abe, 1991a; [5] Idem, 1990; [6] Idem, 1992a; [7] Idem, 1991a; [8] Hosono et al., 1994; [9] Hosono and Abe, 1994; [10] Kasuga et al., 1997; [11] Yamamoto et al., 1997; [12] Yamamoto and Abe, 2000; [13] Idem, 1998; [14] Kasuga et al., 1999b; [15] Hosono and Abe, 1992b; [16] Hosono et al., 1993b; [17] Kasuga et al., 1998; [18] Hosono et al., 1990; [19] Jpn. Pat. Appl. #4-26573 purpose. By utilizing the materials design concept in the present chapter, novel biomedical materials are also developed (Kasuga et al., 1999a).

References

Arberti, G. Acc. Chem. Res. 11:163 (1978)
Alberti, G., U. Constantino. J. Inorg. Nucl. Chem. 40:1113 (1978)
Aono, H., E. Sugimoto, Y. Sadaoka, N. Imanaka, G. Adachi. J. Electrochem. Soc. 136:590 (1989)
Aso, I., M. Nakao, N. Yamazoe, T. Seiyama. J. Catal. 57:287 (1979)
Beall, G. H., D. A. Duke. J. Mater. Sci. 4:340 (1969)
Borisov, B. F., E. V. Charnaya, Y. Kumzerov, A. K. Radzhabov, A. V. Shelyapin. Solid State Commun. 92:531 (1994)
Chibata, I. Immobilized Catal. Kodansha, Tokyo, 97 (1989)
Chow, L. C. J. Ceram. Soc. Japan, 99:954 (1991)
Clearfield, A. *Inorganic Ion Exchanging Materials*. CRC Press, Boca Raton, FL (1982)
Clearfield A., G. D. Smith. Inorg. Chem. 8:431 (1969)
Clearfield, A., R. A. Hunter. J. Inorg. Nucl. Chem. 38:1085 (1976)
Clearfield, A., G. H. Nancollas, R. H. Blessing. *Ion Exchange and Solvent Extraction*, Vol. 5. Marcel Decker, New York, Chap 1(1973)
Clearfield, A., B. D. Roberts. Inorg. Chem. 27:3237 (1988)
Cotton, F. A., G. Wilkinson, L. Paul Gaus. *Basic Inorganic Chemistry*, 2nd ed. Wiley, New York, p. 508 (1976)
Cotton, F. A., G. Wilkinson. Advanced Inorganic Chemistry; In: *A Comprehensive Text*, 4th ed. Wiley, New York, p. 800 (1981)
Doremus, R. H. J. Mater. Sci. 27:285 (1992)
Goetz, A., J. Am. Water Works Assoc. 35:579 (1943)
Goodenough, J. B., H. Y-P. Hong, J. A. Kafalas. Mater. Res. Bull. 11:203 (1976)
Hagman, L. O., P. Kierkegaard. Acta Chem. Scand. 22:1822 (1968)
Haller, W. Nature. 206:693 (1965)
Hammel, J. J. US Patent 3843341 (1974)
Hood, H. P., M. E. Nordberg. U. S. Patent 2106774 (1934)
Hosono, H., Y. Abe. J. Electrochem. Soc. 137:3149 (1990)
Hosono, H., Y. Abe. Solid State Ionics 44:293 (1991a)
Hosono, H., Y. Abe. Phosphorus Res. Bull. 1:351 (1991b)
Hosono, H., Y. Abe. J. Am. Ceram. Soc. 75:2862 (1992a)
Hosono, H., Y. Abe. J. Non-Cryst. Solids 139:86 (1992b)
Hosono, H., Y. Abe. *Nucleation and Crystallization in Liquids and Glasses. Ceramic Transaction*, Vol. 30. The American Ceramic Society, OH, p. 277 (1993)
Hosono, H., Y. Abe. Mater. Res. Bull.. 29:1157 (1994)
Hosono, H., Y. Abe. J. Non-Cryst. Solids 190:185 (1995)
Hosono, H., Y. Abe. *Proc. 18th In. Cong. Glass.* American Ceramic Society,

OH, C2:p. 17 – 26 (1998)
Hosono, H., Z. Zhang, Y. Abe. J. Am. Ceram. Soc. 72:1587 (1989)
Hosono, H., Y. Sakai, M. Fasano, Y. Abe. J. Am. Ceram. Soc. 73:2536 (1990)
Hosono, H., Y. Sakai, Y. Abe. J. Non – Cryst. Solids 139:90 (1992)
Hosono, H., K. Imai, Y. Abe. J. Electrochem. Soc. 140:L7 (1993a)
Hosono, H., K. Imai, Y. Abe. J. Non – Cryst. Solids 162:287 (1993b)
Hosono, H., F. Tsuchitani, K. Imai, Y. Abe. J. Mater. Res. 9:755 (1994)
Kamiya, K., T. Yoko, S. Sakka. J. Non – Cryst. Solids 80:405 (1986)
Kanazawa, T. *Inorganic Phosphate Materials*. Elsevier, Amsterdam, Chap. 6 and 7 (1990)
Kasuga, T., H. Kume, M. Nogami, Y. Abe. Phosphorus Res. Bull. 6:253 (1996)
Kasuga, T., H. Kume, Y. Abe. J. Am. Ceram. Soc. 80:777 (1997)
Kasuga, T., H. Nakamura, K. Yamamoto, M. Nogami, Y. Abe. Chem. Mater. 10:3562 (1998)
Kasuga, T., M. Nogami, Y. Abe. J. Am. Ceram. Soc. 82:765 (1999a)
Kasuga, T., K. Yamamoto, T. Tsuzuki, M. Nogami, Y. Abe. Mater. Res. Bull. 34:1595 (1999b)
Kishioka, A. Bull. Chem. Soc. Jpn. 51:2559 (1978)
Korai, H. J. Antibact. Antifung. Agents 24:509 (1996)
Lee, W-H., H. Kondo, H. Hosono, Y. Abe. Jpn. J. Appl. Phys. 32:1082 (1993)
Mbandza, A., E. Boreds, P. Courtine. Mater. Res. Bull. 20:251 (1985)
Mentre, O., F. Abraham, B. Deffontaines, P. Vast. Solid State Ionics 72:239 (1994)
Messing, R. A.. J. Non – Cryst. Solids 26:482 (1977)
Monma, H., T. Kanazawa. J. Ceram. Soc. Jpn, 84:209 (1976)
Moya, L.. Inorg. Nucl. Chem. Lett. 15:207 (1979)
Nakanishi, K., N. Soga. J. Am. Ceram. Soc. 74:2518 (1991)
Olazcuaga, R., G. Le Flem, A. Boireau, J. L. Soubeyroux. Adv. Mater. Res. 1 – 2:177 (1994)
Orazio, F. D., J. C. Tarczon, W. P. Halpeerin, K. Eguchi, T. Mizusaki. J. Appl. Phys. 65:742 (1989)
Oudet, F., A. Vejux, T. Kompany, E. Boreds, P. Courtine. Mater. Res. Bull. 24:561 (1989)
Qureshi, M., K. G. Varshney. *Inorganic Ion Exchangers in Analytical Chemistry*. CRC Press, Boca Raton, FL (1990)
Rabinovich, E. M., M. I. Shalom, A. Kisilev. J. Mater. Sci. 15:2027, 2039 (1981)
Rawson, H. *Inorganic Glass-Forming Systems*. Academic Press, London (1967)
Res, M. A., J. Bendnarik, J. T. Fourie, J. L. Albain. J. Am. Ceram. Soc. 67:c – 264 (1984)
Roy, R., D. K. Agrawal, J. Alamo, R. A. Roy. Mater. Res. Bull. 19:471 (1984)
Russel, A. D., W. B. Hugo. *Progress in Medical Chemistry*, Vol. 31. Elsevier,

New York, p. 351 (1994)
Schmid, H. K., M. A. Res, F. Blum, H. D. H. Schonberger. Glastech. Ber. 62:64 (1988)
Serghini, A., R. Brochu, M. Ziyad, J. C. Vedrine. J. Alloys Comp. 188:60 (1992)
Simonetti, N., G. Simonetti, F. Bougnol, M. Scalzo. Appl. Environ. Microbiol. 58:3834 (1992)
Stookey, S. D. *Journey to the Center of the Crystal Ball*. The American Ceramic Society, OH (1985)
Subramanian, M. A., R. Subramanian, A. Clearfield. Solid State Ionics 18&19:562 (1986)
Suzuki, T., S. Goto, T. Tanaka. Denkikagaku 52:272 (1984)
Suzuki, T., M. Toriyama, H. Hosono, Y. Abe. J. Ferment. Bioeng. 72:384 (1991)
Sweet, D. V. Registry of Toxic Effects of Chemical Substances 1985 – 1986 Edition. U. S. Department of Health and Human Services. Washington, DC (1985)
Takayama, M. J. Antibact. Antifung. Agents 24:561 (1996)
Tsuhako, M., Y. Horii, K. Kawamoto, N. Kawataka, H. Nariai, I. Motooka. Nippon Kagaku – Kai – Shi (in Japanese) 1810 (1988)
Uo, M., M. Numata, M. Suzuki, I. Karube, A. Makishima. J. Ceram. Soc. Jpn. 100:430 (1992)
Volf, M. B. *Technical Glasses*. Pitman, London, p. 176 (1961)
Warner, T. E., W. Milius, J. Maier. Ber. Bunsenges. Phys. Chem. 96:1607 (1992)
Weetal, H. H. Nature 223:959 (1969a)
Weetal, H. H. Science 166:615 (1969b)
Yamanaka, S. *Inorganic Phosphate Materials*. Elsevier, Amsterdam, Chap. 6 (1989)
Yamamoto, K., Y. Abe. J. Am. Ceram. Soc. 81:2201 (1998)
Yamamoto, K., Y. Abe. Mater. Res. Bull. 35:211 (2000)
Yamamoto, K., T. Kasuga, Y. Abe. J. Am. Ceram. Soc. 80:822 (1997)
Yazawa, T., H. Tanaka. *Porous Materials*. The American Ceramic Society, OH, p. 213 (1993)

5 Tissue Engineering

Qing Liu

5.1 Introduction

Thousands of American people are admitted to the hospitals everyday due to organ failure and tissue loss. It is estimated that over 8 million surgical procedures are performed annually in the United States to treat the millions of Americans who experience organ failure or tissue loss. Physicians treat these patients by transplanting organs from one individual to another, performing reconstructive surgery, or using lifesaving mechanical devices such as kidney dialyzers and mechanical heart valves. However, many of the patients who need organ transplantation (like heart, liver, and kidney transplantations) will die because of the lack of suitable transplant organs. Fortunately, through the developments in the new field of tissue engineering, patients who need new vital tissue and organs have new hope.

Tissue engineering is formally defined as "the application of the principles and methods of engineering and the life sciences toward the fundamental understanding of structure – function relationships in normal and pathological mammalian tissues and the development of biological substitutes that restore, maintain, or improve tissue function" (Skalak and Fox, 1998). More specifically, the term tissue engineering implies some combination of living cells, scaffold materials, and bioactive peptides used to guide the repair or formation of tissue. This form of therapy differs from standard drug therapy in that the engineered tissue becomes integrated within the patient, affording a potentially permanent and specific cure of the disease state. Nowadays, tissue engineering researchers are attempting to engineer virtually every human tissue, like skin, heart valves, teeth, bladders, arteries and veins, cartilage and bone, peripheral and central nervous system tissues, muscles, liver, and pancreatic islet (insulin-producing) cells. Tissue engineering products have such a huge potential market that a lot of capital has been invested in R&D. Table 5.1 summarizes the companies which are currently engaged in tissue engineering R&D and their targeted tissues and organs.

In this chapter, the basic principles of tissue engineering and biomaterials used in tissue engineering will be reviewed. Some important advances in the following applications will be highlighted: tissue engineering of skin, bone (skeletal system), peripheral nerve (nerve system), and the heart valve (vascular system).

Table 5.1 Human tissue engineered through tissue engineering

Name of Tissue or Organ	Name of Company
Bone	Curis (Cambridge, MA); Orquest (Mountain View, CA); Sulzer Orthopedics Biologics (Austin, TX); Genetics Institute (Cambridge, MA); Osiris Therapeutics (Baltimore, MD); Regeneron (Tarrytown, NY); IsoTis (Bilthoven, The Netherlands).
Cartilage	Genzyme Tissue(Cambridge, MA); Biomatrix (Ridgefield, NJ); Integra LifeSciences (Plainsboro, NJ); Advanced Tissue Sciences (La Jolla, CA); ReGen Biologics (Franklin Lakes, NJ); Osiris Therapeutics; IsoTis (Bilthoven, The Netherlands).
Skin	Organogenesis (Canton, MA); Advanced Tissue Sciences (La Jolla, CA); Integra LifeSciences; LifeCell (Woodlands, Texas); Ortec International (New York, NY); IsoTis.
Heart valves, arteries, and veins	Organogenesis; Advanced Tissue Sciences; Genentech; LifeCell; Reprogenesis (Cambridge, MA).
Pancreas	BioHybrid Technologies (Shrewsbury, MA); Neocrin (Irvine, CA); Circe Biomedical (Lexington, MA).
Urinary tract	Reprogenesis (Dallas, TX); Integra LifeSciences.
Teeth	Biora (Chicago, IL); Atrix Laboratories (Fort Collins, CO.); Curis.
Breast	Reprogenesis; Intergra LifeSciences.
Liver	Advanced Tissue Sciences; Organogenesis.
Bladder	Reprogenesis.
Spinal cord nerves	Acorda (Hawthorne, NY); Regeneron (Tarrytown, NY); StemCells (Sunnyvale, CA); Guilford Pharmaceuticals (Baltimore, MD).

5.2 Tissue Engineering Approaches

Illustrated below is the principle of tissue engineering. To create a living replacement tissue, two basic approaches can be used, depending on the targeted tissue (Fig. 5.1).

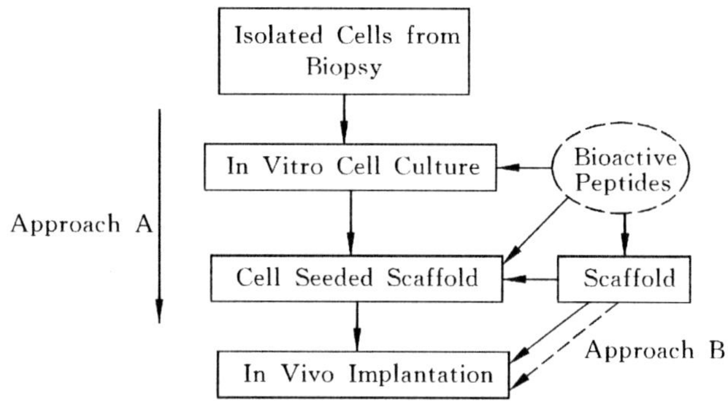

Figure 5.1 Tissue engineering approaches in regeneration of tissue or organ

In approach A, a small number of cells can be harvested from the patient using a biopsy, and then the harvested cells are cultured *in vitro* in order to get the appropriate numbers in the laboratory. These cells can then be grown within a three-dimensional natural or synthetic scaffold and in the presence of appropriate growth and differentiation factors. If provided with the appropriate conditions and signals, the cells will secrete various matrix materials to create an actual living tissue that can be used as a replacement tissue to be implanted back into the defective site in the patient. The scaffold should ultimately degrade to prevent certain risks that can occur with the long-term presence of any foreign material in the body. If cells from the patient are used, then there will be no immune rejection response to the implanted tissue. In approach B, scaffold materials loaded with or without appropriate growth and differentiation factors can be implanted into the defect site, the scaffold materials will guide the tissue growth *in situ* with the help of appropriate growth and differentiation factors secreted by the host or by the release of loaded factors.

Depending on the regeneration location, tissue engineering methods can also be classified as follows:

(1) *In vitro* regeneration The new tissue is designed and grown outside the body for later implantation to repair or replace diseased tissues. The most common example of this form of therapy is the skin graft which is used for treatment of burns.

(2) *In vivo* regeneration New tissue is generated within the body through local delivery of cells, growth factors, genes, or simply by the presence of a scaffold which guides tissue regeneration. Examples are bone cell delivery in

bone tissue engineering and peripheral nerve regeneration using nerve guidance channels.

(3) Partial regeneration Only part of the tissue is regenerated either *in vitro* or *in vivo* to replace the function of diseased internal tissues. This approach involves isolating cells from the body, placing them on or within structural matrices, and implanting the new system inside the body or using the system outside the body. An example of this approach is the tissue engineered heart valve.

5.3 Cells Used in Tissue Engineering

In general, cells which can form replacement tissue will be used to construct tissue engineering constructs, because the cellular component is ultimately responsible for the performance of the replacement tissue in regard to its function. The tissue engineering process may begin with the identification of the relevant cell type and the isolation of these cells from the host tissue. Then the next step is obtaining enough cells from the available host tissue. This step is called expansion.

Expansion of the cell population *in vitro* is a critical step in the process of tissue engineering. Some cell types multiply easily *in vitro*, like fibroblasts which are used for skin and tendon regeneration. Some cell types are diffcult to multiply, like neurons in nerve tissue regeneration. Some cells dedifferentiate upon repeated passing, like chondrocytes in cartilage regeneration. It has been shown that when chondrocytes were cultured *in vitro* in two-dimensional petri dishes for a long period, the collagen synthesized by the chondrocytes switched from type Ⅱ to type Ⅰ, and the appearance of chondrocytes became fibroblast-like (Rodriguez and Vacanti, 1998).

In some cases, a limited cell source is the main problem. Recently, it has been found that stem cells are able to generate virtually all other cell types when using the appropriate differentiation factors. For example, mesenchymal stem cells are capable of differentiating into bone, cartilage, tendon, and muscle (Caplan, 1990). Hematopoietic stem cells can give rise to bone (Lazarus et al., 1995). Neural stem cells can give rise to neurons, astrocytes, and oligodendrocytes (Hulspas et al., 1997). More recently, human embryonic stem cells have been found able to give rise to essentially all cell types in the body (Thomson et al., 1998). For example, they can give rise to heart muscle cells, pancreas islets, and cartilage, etc. when different differentiation factors are added. The use of differentiation factors is critical to regulate lineage formation and tissue development.

Growth factors are commonly used to support the various terminal phenotypes (Nimni, 1997). For example, transforming growth factor-β (TGF-β) has been used to support chondrocyte growth. Bone morphogenetic proteins

(BMP) have been used to induce *in vivo* bone formation. The basic fibroblast growth factor (bFGF) has been used to stimulate nerve tissue regeneration.

5.4 Biomaterials for Tissue Engineering

Biomaterials are commonly used in tissue engineering both as scaffolds and carriers of cells, growth factors, and genes. Virtually all types of biomaterials, i. e. polymers, metals, ceramics, and composites, are used in tissue engineering. A scaffold should provide a necessary mechanical support as well as a physical structure for the transplanted cells to attach, grow, and maintain differentiated functions. As carriers for delivery cells and bioactive substances, the materials should possess the necessary surface chemistry for cell attachment and proliferation and for biological substance immobilization.

Tissue engineering often prefers to use biodegradable biomaterial because as the new tissue develops, the new tissue will gradually replace the scaffold. But in some cases, the currently available biodegradable biomaterials cannot meet clinical needs in terms of their mechanical properties; therefore, nonbiodegradable biomaterials have to be used.

5.4.1 Polymeric Biomaterials

Polymeric biomaterials include synthetic and naturally occurring polymers. They are playing critical roles in tissue engineering.

Polymeric biomaterials have physical properties which closely resemble those of soft tissue. Therefore, they are extensively used as implant materials and scaffold materials. Polymers are relatively easy to manufacture into different shapes and structures. Polymers are also easy to chemically modify or functionalize via chemical and biochemical reactions. Polymers also can be made biodegradable, which is considered crucial in tissue engineering. Best of all, with the progress in polymer science, it is possible to tailor the physicochemical properties and structures of polymers via molecular design.

Polymeric biomaterials can be used directly as scaffolds without any cells or bioactive compounds previously incorporated to direct cell ingrowth into specific architectures or act as a barrier to prevent undesired scarring tissue growth. When polymeric biomaterials are preloaded with bioactive substances in bulk or on the surface, the biomaterials have the ability to regulate the tissue growth or a regenerative process.

Following are the commonly used polymers in tissue engineering.

5.4.1.1 Natural Polymers

1. Collagen

Collagen is the major component of mammalian connective tissue. It is found in every major tissue that requires strength and flexibility, such as tendons, skin, and fascia. There have been 14 types of collagen identified to date, among which type I collagen is the most abundant form.

Collagen-based medical devices have been used since the turn of this century. Gut sutures and porcine skin are a few examples. Nowadays, there are numerous collagen-based medical devices approved by the FDA.

Collagen proteins are characterized by a unique triple helix formation extending over a large portion of the molecule. Every third amino acid is glycine, and about 25% of the amino acid residues are proline and hydroxyproline. The three peptide subunits of type I collagen have similar amino acid compositions, with each chain composed of about 1050 amino acid residues. The length of each subunit is about 300 nm, and the diameter of the triple helix is about 1.5 nm (Pachence, 1996).

When collagen is implanted *in vivo*, it is subjected to degradation attacks by collagenases. It also frequently elicits an immune response of the host if the collagen is from an animal source. Chemical modification of collagen (or cross-linking) will make it less susceptible to enzymatic degradation and also reduces its immunogenicity.

Collagen can be cross-linked by using chemical agents, such as formaldehyde or glutaraldehyde, or by physical techniques, such as dehydrothermal treatment at (110℃ in a vacuum) (Weadock et al., 1984). The chemical cross-linking of collagen using aldehyde involves the reaction between the aldehyde group and an ε-amino group of lysyl residues in collagen, which forms interchain cross-links. Dehydrothermal treatment of collagen will promote the condensation of carboxylic groups and amino groups in collagen by removing the aqueous product of condensation. This condensation reaction leads to the formation of interchain amide links.

When used in tissue engineering as a scaffolds, collagen has to be manufactured into a three-dimensional scaffold. A porous structure provides three critical functions for the scaffold. First, pore channels allow the migration or entry of cultured cells to the scaffold. Second, the porous structure gives the scaffold an enormous specific surface area for cells to seed and interact with the scaffold. Third, the porous structure allows nutrients to diffuse into the scaffold to support the growth of the seeded cells.

Porous collagen scaffolds can be prepared by using a freeze and dry method. In this method, a diluted collagen fiber suspension is first frozen at a low temperature and then the frozen suspension is exposed to a low temperature vacuum to induce sublimation of the ice crystals. The formed pore morphology and structure of the pores can be controlled by the conditions of ice nucleation and growth, i.e. the freezing temperature and freezing time (Dagalaskis et al., 1980).

Collagen scaffolds have been used in the tissue engineering field to engineer many tissues, including bone, cartilage, nerve, skin, blood vessels, etc.

2. Glycosaminoglycans (GAGs)

GAGs are a group of natural occurring materials. Hyaluronic acid, chondroitin sulfate, dermatan sulfate, keratan sulfate, and chitin are frequently seen examples of GAGs. The structure of GAGs can be generally described as that of an alternating copolymer consisting of repeated units, a hexosamine (glucosamine or galactosamine), and a sugar (galactose, glucuronic acid, or iduronic acid) (Yannas, 1996).

Among the GAGs, hyaluronic acid can be used to prepare hydrogels for use in ophthalmology. Chondroitin-6-sulfate can induce precipitation of collagen at an acidic pH. The formed coprecipitates can be subsequently dried and covalently cross-linked to yield a graft copolymer of collagen and glycosaminoglycans. This material is used in the regeneration of skin, nerves, and the meniscus of joints (Yannas, 1996).

5.4.1.2 Synthetic Polymers

1. Poly(α-hydroxy acid)

Poly(α-hydroxy acid) is a general name for a class of polymers including poly (lactic acid) (also known as polylactide), poly (glycolic acid), and their copolymers (Fig. 5.2).

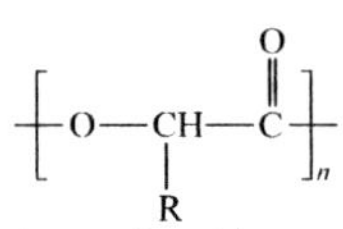

Poly (Lactic Acid) ($R=CH_3$);
Poly (Glycolic Acid) (R=H)

Figure 5.2 The structure of poly(lactic acid) and poly(glycolic acid)

These polymers are currently the most widely investigated and most commonly used synthetic biodegradable polymers in the biomaterials and tissue engineering fields. These polymers are also among some of the few biodegradable materials that have been approved for human use in various medical devices. One of the important reasons for the wide use of these polymers is that the constituent units of these polymers are derived from natural metabolites.

Lactic acid is a chiral molecule, existing in two stereoisometric forms: D-LA and L-LA. Therefore, it gives three corresponding polymers: two stereoregular polymers poly (D-LA) and poly (L-LA), and the racemic form poly(DL-LA). Both poly (D-LA) (PDLA) and poly (L-LA) (PLLA) are crystalline polymers, while poly(DL-LA) (PDLA) is an amorphous polymer. PLLA is the most commonly used form, as opposed to PDLA, because the degradation product L-lactic acid is the natural occurring stereoisomer of lactic acid (Kohn and Langer, 1996)

As a biodegradable polymer, PLLA has satisfactory *in vitro* biocompatibility (van Sliedregt et al., 1992, 1993). It is essentially non-toxic and elicits only a mild inflammatory response. The hydrolysis product L-lactic acid is the normal intermediate of carbohydrate metabolism and will not accumulate in vital organs. It has been proposed and successfully applied to the reconstruction of bone, articular defects, suture materials, drug carriers, and fixation devices (Verheyen et al., 1992a).

PLLA degrades slowly *in vivo*. It has been found that when bone plates made of crystalline (PLLA) were used for the fixation of zygoma fractures, after 3 years, all of the patients showed a severe, well-defined swelling, strictly limited to the implantation site. In the case of seven patients, surgeries were performed to surgically remove the swelling. During light microscopic examination, the removed tissue was characterized by a foreign body reaction without signs of inflammation. At the transmission electron microscopic level, large amounts of highly crystalline PLLA particles were observed. It was concluded that the degradation of PLLA was very slow and that the total degradation time was much longer than 3 years (Rozema et al., 1992).

To overcome this problem, D-lactic acid was added to obtain a biomaterial with a lower crystallinity than PLLA and consequently a higher degradation rate.

Poly(glycolic acid) (PGA) is a highly crystalline linear aliphatic polyester. It has been used in suture materials and bone fixation devices. PGA degrades relatively fast compared to PLLA. Therefore, PGA was frequently used as the comonomer to make a copolymer with PLLA to control the degradation rate of the copolymer. A commercially available PLA/PGA (PLGA, 10%/90%) plate for bone fixation under the names SR – PGA and Biofix have been used worldwide in over 20, 000 patients (Dunn, 1995).

When used in tissue engineering as a scaffold, poly(α-hydroxy acid) has to be made into three-dimensional scaffold. There have been many processing methods developed to prepare PLA, PGA, and their copolymer scaffolds. One of the commonly used methods is called solvent-casting and the particulate-leaching technique. In this technique, the polymer was first dissolved in a solvent, such as chloroform and methylene chloride. Salt particles of a particular size, such as sieved NaCl particles, were dispersed in the polymer solution. Then the dispersion was cast in a glass container to allow solvent to evaporate. After evaporation of the solvent, a composite sheet of polymer – salt particles was obtained. The composite membrane then could be further heated above the melting temperature of the polymer. After cooling, the membrane was immersed in water to leach out the salt, and the resulting porous polymer structure was dried under vacuum conditions. Highly porous PLLA foam with a porosity up to 93% was prepared using this technique. The advantages of this technique are that the porosity of the foam can be controlled by changing the ratio of salt to polymer, the pore size can be controlled by changing the size of the salt particles, and the crystallinity of the foam can be controlled by the cooling rate of the mold. However, there is a limit for this method, that is, this technique can only be used to manufacture membranes not more than 2 mm in thickness (Lu and Mikos, 1996; Widmer and Mikos, 1998a).

In order to overcome this drawback, many other techniques were developed to manufacture three-dimensional porous structures with certain shapes. A technique called the lamination technique was used to join the composite membranes obtained using the above mentioned technique by wetting the surface of membranes with a solvent prior to lamination. The laminated membrane can then be shaped into the desired shape simply by cutting it. In another method called melt molding, composite membranes obtained by solvent casting, as described above, were cut into pieces and put into a mold. The mold was then heated above the glass transition temperature of the polymer, and pressure was applied to the mixture in order to bond the composite pieces together. After cooling down, the salt particles were leached out to get a porous structure. An extrusion technique can also be applied to the polymer – salt particle membrane to manufacture a porous scaffold with a more complicated shape (Lu and Mikos, 1996; Widmer and Mikos, 1998a).

It has to be mentioned that the above mentioned techniques can also be applied to many other polymers. Many other techniques, such as fiber bonding, phase separation (Zhang and Ma, 1999; Liu et al., 2000), three-dimensional printing, and a high-pressure process were also developed to manufacture the various polymer scaffolds. Readers can refer to two excellent reviews (Lu and Mikos, 1996; Widmer and Mikos, 1998a).

Recently, we developed a method to apply a porous PLGA coating onto polymethylmethacrylate (PMMA) disks as craniofacial prostheses. In this coating method, PLGA was first dissolved in chloroform. Then the PMMA disk was dipped into a PLDA solution. A layer of NaCl particles adhered to the PLGA coating before the coating completely dried out. After repeating the dipping-adhesion process several times, a composite coating of PLGA and salt particles was obtained. The salt particles were leached out in distilled water, and a porous PLGA coating was then obtained (Fig. 5.3) (Dean et al., 1999).

2. Polyactive™

Polyactive™ is the trade name of a series block copolymers of polyethylene glycol (PEG) and polybutylene terephathalate (PBT) (Fig. 5.4). As a biomaterial, it has been extensively studied since the discovery of its bone – bonding properties in 1989 (Bakker et al., 1989; van Blitterswijk et al., 1992).

In this copolymer, PEG serves as the soft segment, and PBT serves as the hard segment. When changing the ratio of PEG/PBT, a series of copolymers can be obtained with different mechanical, physicochemical, and biological properties.

In addition to the change in the PEG/PBT ratio, the molecular weight of PEG can also be changed, which leads to an additional change in the mechanical

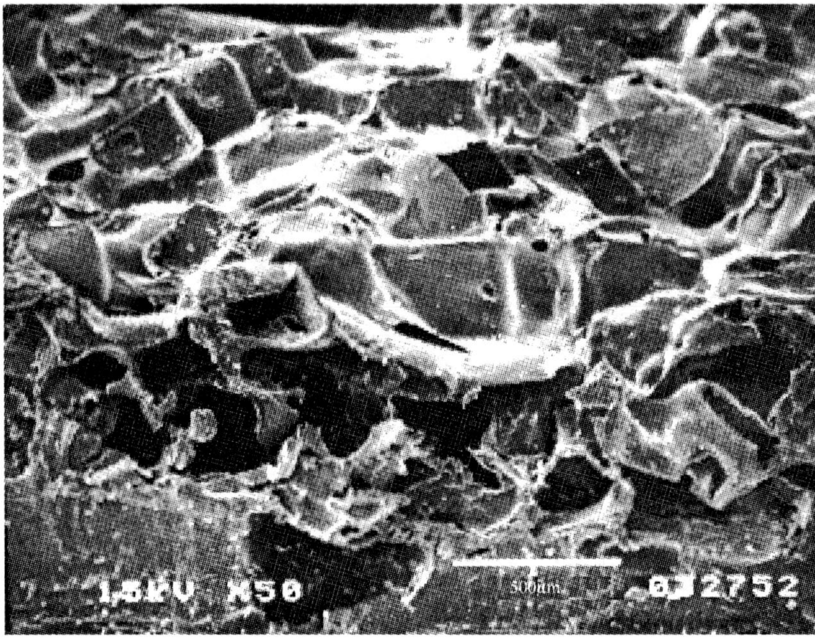

Figure 5.3 A scanning electron microscopic picture showing the cross section of a porous PLGA coating (top) on a PMMA disc (bottom)

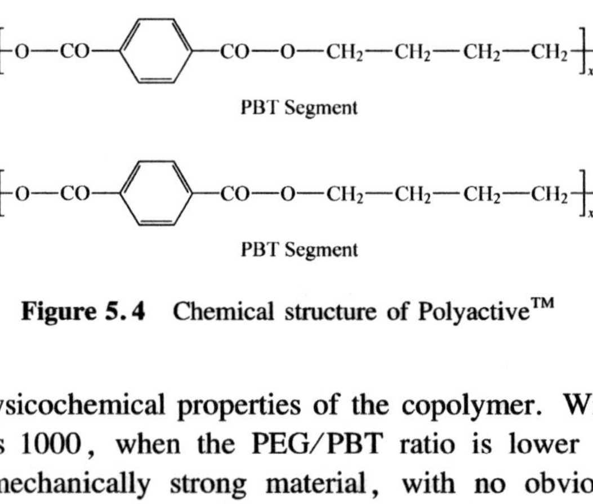

Figure 5.4 Chemical structure of Polyactive™

properties and physicochemical properties of the copolymer. With the molecular weight of PEG as 1000, when the PEG/PBT ratio is lower than 55/45, the copolymer is a mechanically strong material, with no obvious water uptake ability. The copolymer becomes more and more rubber-like when the PEG content is increased. It has also been found that when the ratio of PEG/PBT is higher than 55/45, the copolymer is a biodegradable polymer and calcifies postoperatively, thereby inducing bone bonding.

Bone bonding is a unique property for Polyactive™. And so far, Polyactive™ is the only polymer which can bond to bone when implanted *in vivo*. Regarding the bone-bonding property, it has been proposed that the PEG segment of Polyactive™ can complex calcium ions from the physiological environment by a similar mechanism proposed for the calcification of polyetherurethane (van Blitterswijk et al., 1992, 1993). The calcification of Polyactive™ *in vivo* is at least composed of carbonated apatite which is similar to the apatite layer formed on other bioactive ceramics and to the mineral phase of bone (Radder et al., 1993, 1994). It is hypothesized that following the calcification of the PEO/PBT copolymer surface, bone apposition may take place in a manner similar to that on the apatite surface layer of other bioactive biomaterials which initially contain calcium and phosphorus ions.

Since Polyactive™ has been shown to be satisfactorily biocompatible, and biodegrable as well as possessing the necessary bone bonding properties, it has been studied extensively as a bone replacement material (Radder et al., 1993, 1994; Liu et al., 1996, 1997, 1998b; Kuijer et al., 1998) as a dental implant materials (Meijer et al., 1995, 1997). It also has been studied for use as a scaffold in skin regeneration (van Dorp et al., 1999; Xiao et al., 1999) and guided bone tissue regeneration (Jansen et al., 1995).

Polyactive™ porous scaffolds can be prepared using solvent casting and the particulate-leaching technique (van Dorp et al., 1999; Xiao et al., 1999).

3. Poly(propylene fumarate)

Poly(propylene fumarate) (PPF) is an alternate copolymer of propylene glycol and fumaric acid (Fig. 5.5).

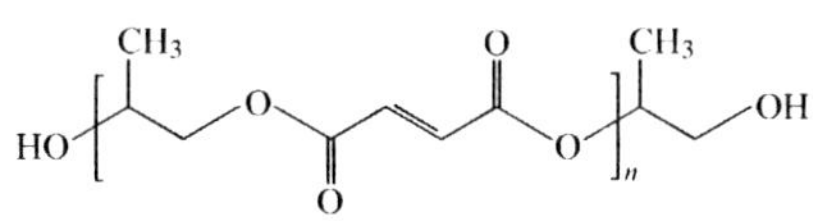

Figure 5.5 Chemical structure of PPF

It has ester linkages in its structural unit which is the basis for its *in vivo* biodegradation through a hydrolysis mechanism. The unsaturated double bond in its structural unit offers the possibility of cross-linking the polymer via radical polymerization. The structural units, propylene glycol and fumaric acid, can be excreted via normal metabolic pathways. Fumaric acid is a component of the Krebs cycle, and propylene glycol has been used as an intravenous fluid component.

This polymer was developed mainly as an injectable bone cement (Peter et al., 1997). The main advantage of this unsaturated polymer is its ability to cure the material *in vivo*, thereby filling in a skeleton defect of any shape or size with minimal surgical intervention. PPF can be cured or cross-linked by using a vinyl monomer with a radical initiation system such as benzoyl peroxide/dimethyl toluidine. Adjusting the PPF to vinyl monomer ratio can change the ultimate mechanical properties and the degradation rate of the cured polymer. Solid phase components can be incorporated into the polymer to form a composite material. These components can either serve as a porogen (in the case of NaCl) or enhance the osteoconductivity and mechanical properties (in the case of β-tricalcium phosphate) (Yaszemski et al., 1995).

In vivo studies have demonstrated that a composite formulation incorporating sodium chloride as a porogen and a β-tricalcium phosphate as an osteoconductive phase elicited a mild inflammatory response when implanted subcutaneously in the back of a rat (Peter et al., 1998b). In the rat proximal tibia model, the studies showed significant bone ingrowth into the polymer at 5 weeks postimplantation.

4. Polyphosphazenes

Polyphosphazenes are a broad class of macromolecules with a highly flexible inorganic phosphorus – nitrogen backbone (Fig. 5.6).

Figure 5.6 Chemical structure of polyphosphazenes

The R side groups can be chosen from a wide variety of organic, organometallic or inorganic units. When changing the side groups, a wide range of polyphosphazenes can be obtained with various properties, from water soluble to water insoluble, rigid to flexible, crystalline to amorphous. Thus, polyphosphazenes are one of the most versatile class of organic/inorganic polymers known.

One of the very promising applications for polyphosphazenes is their use in biomaterials (Allcock et al., 1988, 1990; Allcock and Chang, 1991; Shriver et al., 1987; Tenhuisen et al., 1996). Degradation of the phosphorus – nitrogen backbone yields the biologically compatible hydrolysis products, phosphate and ammonia. Even though most polyphosphazenes are stable to moisture, considerable effort has been made on the synthesis of polymers which degrade at a biological pH and yield innocuous degradation products (Allcock et al., 1990; Crommen et al., 1992a, 1992b; Domb et al., 1992).

Polyphosphazenes have been used as scaffolding materials for bone tissue engineering. Studies have been performed on the growth, proliferation, and phenotypic expression of osteoblasts on polyphosphazenes bearing amino acid ester side groups. *In vivo* work showed that this biomaterial elicited minimal inflammatory response and is capable of supporting bone growth (Laurencin et al., 1998a). Recently, calcium cross-linked polyphosphazene has showed its promise in bone tissue engineering applications (Tenhuisen et al., 1996).

Polyphosphazenes can be processed into three-dimensional scaffolds using cast and salt leaching techniques (Laurencin et al., 1996b).

5. Polypyrrole

Polypyrrole (PP) (Fig. 5.7) is a conductive polymer. In its oxidized form, polypyrrole is a polycation, having delocalized positive charges along its highly conjugated backbone, and is an electronic conductor (Prezyna et al., 1991). Its charge neutrality is achieved by the incorporation of negatively charged ions termed "dopants". PP also exhibits reversible electrochemistry allowing for the control of the surface-charge density by varying the oxidation state of the polymer.

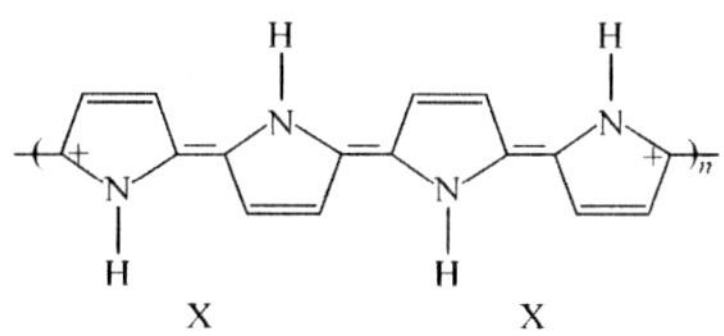

Figure 5.7 Chemical structure of polypyrrole

Unlike piezoelectric materials, PP does not require mechanical deformation for the generation of electrical charges. An electric current can pass through PP. Therefore, PP can be used to deliver local stimulation to cells and tissue. Amongst all the conducting polymers, polypyrrole was chosen for study as a biomaterial because it is easy to prepare by standard electrochemical techniques and its surface charge characteristics can be modified easily by changing the dopant anion (X^-) that is incorporated into the material during synthesis.

By choosing an appropriate dopant anion, the properties of PP can be tailored to a specific application. For example, one of the most interesting properties of PP is that not only small anions can be used as dopants, but also polymeric anions can be used as dopants in PP. Therefore, it is possible to incorporate bioactive macromolecules into PP. It has been shown that collagen (Li and Khor, 1995), heparin (Garner et al., 1999), enzymes (Shin and Kim, 1996), and hyaluronic acid (Collier et al., 2000) can be incorporated to PP as dopants.

The surface characteristics of PP such as charge density and wettability can be changed by the control of the oxidation state or by the incorporation of dopants. It has been shown that the charge density and wettability of PP can be reversibly changed with an applied electric potential, which could provide a noninvasive means to control the shape and function of adherent cells.

Potential biomedical applications for PP have been developed based on its electrochemical properties. These include the use of PP as a matrix for the controlled delivery of dopamine (Miller and Zhou, 1987), as a biosensor for detection of glucose (Couves, 1989) or proteins (Sadik and Wallace, 1993), and as a nerve guidance channel for the regeneration of peripheral nerves (Shastri et al., 1996; Schmidt et al., 1997; Furnish and Schmidt, 1998).

5.4.2 Metals

Metallic implants have been frequently used as bone replacement materials, either as permanent prostheses such as hip prostheses, dental implants, etc, or as temporary implants such as plates, pins, screws, and rods for the fixation of bone fractures. They are chosen as biomaterials simply because they have superior mechanical properties which allow them to be used in load-bearing situations. Currently, stainless steel (Small and Misiek, 1986), cobalt – chrome alloys (Albrektsson et al., 1986), and titanium and its alloys (Albrektsson et al., 1986; Steflik et al., 1993) are used to fabricate implants. These implants are usually not integrated by bone tissue or are integrated only after extended implantation periods. Improvements in the implant integration of bone can be accomplished by cement fixation, use of a porous bead implant surface to allow bone ingrowth to achieve a mechanical fixation, or the application of bioactive ceramic coatings such as hydroxyapatite coatings. Hydroxyapatite coatings applied by various methods (de Groot et al., 1987; Ducheyne et al., 1980; Lacefield, 1988; van Raemdonck et al., 1984) are in clinical use today and improve implant performance. Recently, tissue engineering methods have been applied to metallic prostheses to improve implant – bone integration. Growth factors and other bioactive molecules can be applied to the surface of metallic implants (Endo, 1995; Lind et al., 1996a, 1996b; Nanci et al., 1998; Peleo, 1995, 1997). Cells can also be cultured on metal implants before implantation (de Bruijn et al., 1999).

On some occasions, the high mechanical stiffness of metallic implants may result in stress shielding and bone resorption due to the mismatch of the elastic modulus of metals with that of bone (Daniels et al., 1990; Terjesen and Apalset, 1988). Other disadvantages of using metallic implants include the need for a second operation to remove temporary implants and the negative tissue response caused by the ions released from permanently implanted devices (Black, 1981; Sinibaldik et al., 1976).

5.4.3 Ceramics

The use of ceramics dates back to the nineteenth century when calcium sulphate (plaster of Paris) was first used by Dreesmaanas as a plaster for the fixation of bone. Nowadays, the commonly used bioceramics are metallic oxides (e. g., Al_2O_3, MgO), calcium phosphate [e. g., hydroxyapatite (HA), tricalcium phosphate (TCP), octacalcium phosphate (OCP)] (de Groot, 1981; LeGeros, 1991), and glass-ceramics (e. g., Bioglass, Ceravital) (Hench et al., 1971; Kokubo et al., 1990). Metallic oxides are considered to be nearly bioinert in biological environments, while calcium phosphate and glass ceramics are bioactive because they can bond to bone in bony sites when implanted. Because of the good biocompatibility and bioactivity of bioceramics, they have been successfully used in hard tissue replacement.

Among the bioactive ceramics, synthetic hydroxyapatite with the chemical composition $Ca_{10}(PO_4)_6(OH)_2$ has been extensively studied as a bone replacement material. As a bulk material, HA lacks sufficient tensile strength and is too brittle to be used in most load-bearing applications. Therefore, when hydroxyapatite has to be applied in load-bearing situations, the material is coated onto a metal core (de Groot et al., 1987) or is incorporated into polymers as composites (Bonfield et al., 1983; Doyle et al., 1991; Liu et al., 1996, 1997, 1998a; Verheyen et al., 1992a).

Various coating methods have been used to coat hydroxyapatite on metal implants. Besides the commonly used plasma spray coating method (de Groot et al., 1987), biomimetic coating methods have attracted much attention. It is believed that the advantages of biomimetic coating methods over other coating methods are (1) conformal coatings can be produced on complex-shaped and/or microporous implants, (2) no adverse effects from the heat on the substrates occur, and (3) the biomimetic coating is more bioactive because it has crystal structure and crystallinity similar to that of bone minerals. It has been shown that calcium phosphate coatings can be quickly formed in a physiological condition (pH 7.4, 37℃) on surface modified titanium and its alloys (Wen et al., 1997, 1998), on modified bamboo (Li et al., 1997), and on carbon/carbon composites (Li et al., 1998).

Due to their excellent biocompatibility and osteoconductivity, porous hydroxyapatite ceramics have been extensively studied in bone tissue engineering (Bruder et al., 1998; Chistolini et al., 1999).

Porous HA scaffolds can be manufactured using foaming agents. In this technique, HA particles are first mixed with a liquid organic binder (e. g., polyvinyl alcohol) and a surfactant to obtain a homogenous mixture. Then the mixture is foamed. After foaming, the formed HA is heated at a temperature which is sufficient to substantially burn out the organic binder. The obtained macroporous ceramic foam is suitable for use in biomedical applications such as synthetic bones, tissue engineering scaffolds, and drug delivery devices (Hing and Bonfield, 2000).

5.4.4 Ceramic/Polymer Composite Biomaterials

The use of particulate fillers to reinforce polymeric biomaterials is quite important and quite successful in clinical applications, such as dental restorative resins and bone cement.

The purpose of using filler particles to reinforce the polymer matrix is to improve the mechanical properties such as the elastic modulus (Guida et al., 1984; Castaldini et al., 1984; Bonfield et al., 1984, 1986) and fatigue behavior

(Castaldini et al., 1987), and to improve the bioactivity or bone-bonding properties (Bonfield et al., 1986; Doyle et al., 1991; Knowles et al., 1992; Verheyen et al., 1993). Some other benefits may also be obtained by using fillers, to diminish the creep of composites (Castaldini et al., 1986) and to decrease the temperature rise during the polymerization of bone cements (Guida et al., 1984).

The use of bioactive fillers such as hydroxyapatite (HA), AW ceramic, or Bioglass particles to reinforce a polymer may improve both the mechanical properties and the bone bonding properties. As indicated by Bonfield et al. (1984, 1986), the elastic modulus of polyethylene (PE) can be increased from 1 GPa to about 8 GPa, which is in the low band of the value for bone, retaining a fracture toughness comparable to bone. When implanted *in vivo*, the HA/PE composites can induce bone apposition and thus create a secure bond between the natural bone and the implant. Inspired by this work, research has been extended to the biodegradable polymer matrix. When implanted *in vivo*, such composites will induce bone formation or bone ingrowth, and as the biodegradable polymer matrix degrades, the implant will finally be replaced by bone tissue. The load thus can be gradually transferred to the newly formed bone. Based upon this idea, several hydroxyapatite reinforced biodegradable polymer composites have been developed, such as HA/polyhydroxybutyrate (Doyle et al., 1991, Knowles et al., 1992; Boeree et al., 1993) and HA/polylactide (Verheyen et al., 1992a, b, 1993). Recently, nanosize hydroxyapatite (nanoapatite) particles have attracted significant attention from researchers since the nanoapatite particles are similar to bone minerals in terms of crystal size and crystallinity (Li et al., 1994, 1997). They may offer a great improvement in the bone-bonding properties of composites. It was found in *in vitro* experiments that adding nanoapatite to Polyactive™ greatly increases the ability of composites to induce calcium phosphate precipitation. As can be seen from Fig. 5. 8, when adding 10 wt% nanoapatite to a Polyactive™ matrix, the composites can induce calcium phosphate precipitation in relatively short period compared to pure Polyactive™ (Liu et al., 1997, 1998c).

The use of a filler to reinforce a biodegradable polymer matrix offers another advantage: the possibility to control the biodegradation rate. It has been shown that by the addition of basic fillers, such as HA and magnesium oxide, the degradation rate as well as the degradation mechanism of poly (DL-lactide) can be changed (van der Meer et al., 1996). Jones and Williams (1996) also showed that the degradation pattern of poly (L-lactide) was affected by the addition of ceramic filler. Although the mechanical properties of composites can be improved to a certain extent by the addition of bioactive filler particles, it has been stated by several researchers (Verheyen et al., 1992a, b; Wang et al., 1994) that there is still a need to improve the bonding between the filler and the matrix since there is clearly no other bonding force between the two phases than mechanical interlock. The use of certain coupling agents was suggested (Verheyen et al., 1992a, b; Wang et al., 1994).

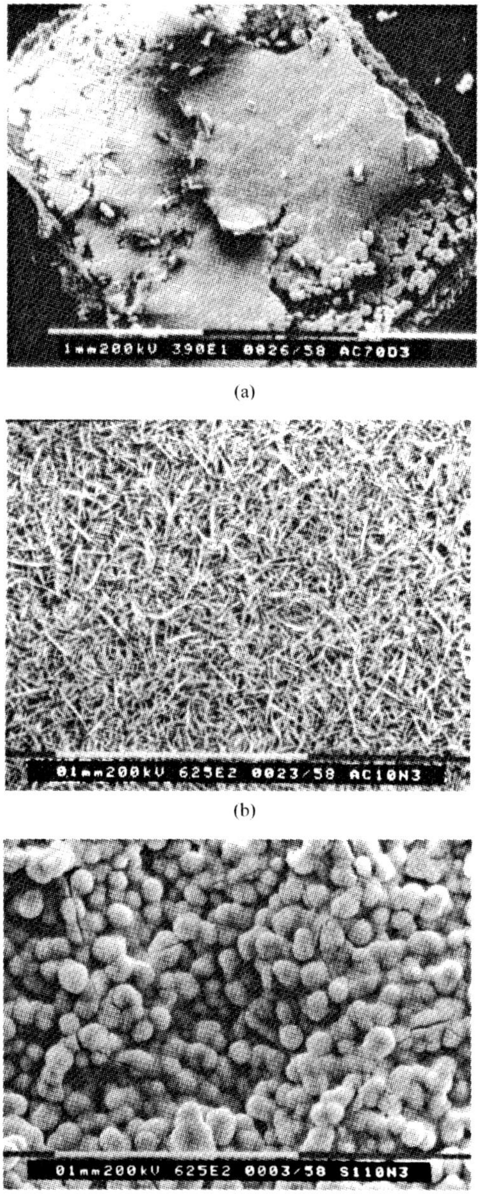

Figure 5.8 Scanning electron microscopic pictures showing the calcification developed on Polyactive™ and its composites. (a) The incomplete coverage of Polyactive™ 70/30 by the calcium phosphate mineral layer after a 3-day immersion in accelerated calcification solution (ACS). (b) The complete coverage of 10% nanoapatite/polymer composites by calcium phosphate crystal plates after a 3-day immersion in ACS. (c) Complete coverage of 10% nanoapatite/polymer composites after a 3-day immersion in simulated body fluid. (Liu et al., Cells Mater. 7:41, 1997)

A coupling agent is an additive which promotes the development of a strong bond between a filler (fiber) surface and a polymer. Silane coupling agents have been widely used to improve the bonding strength of the two phases and have the general formula of

$$X_3SiRY$$

X represents a hydrolyzable group and Y is an organofunctional group. The organofunctional groups are chosen for their reactivity or compatibility with the polymer, while the hydrolyzable groups are merely intermediates in the formation of a bond with the filler or fiber surface. The exact behavior of the coupling agent is a matter of some controversy.

Silane coupling agents in mineral filled dental resins have been employed since the 1960s (Bowen, 1963; Venhoven et al., 1994; Jones and Rizkalla, 1996) as a method to improve the bonding of the filler to the resin. It was reported that appropriate silane coupling agents were chosen for a variety of mineral fillers to improve the mechanical properties of the composites (Plueddemann, 1991). The greatest improvement was observed with silica, alumina, glass, silicon carbide, and aluminum needles. A good but somewhat lesser response was observed with talc, wollastonite, iron powder, clay, and hydrated aluminum oxide, while only a slight improvement was imparted by asbestine, hydroxyapatite, titanium dioxide, and zinc oxide. Surfaces that showed little or no apparent response to silane coupling agents included calcium carbonate, graphite, and boron. Those results suggest that the coupling activity of silanes is not universal to all mineral surfaces.

The effect of the use of silane coupling agents on hydroxyapatite (HA) seems to have somewhat controversial results. It has been shown that silane treated HA particles have a positive effect on the mechanical properties of a composite. Behiri et al. (1991) showed that applying methacryloxy-propyltrimethoxysilane (MPS) to the surface of HA particles may enhance the tensile modulus, yield stress, and elongation to the fracturing of polyethylmethacrylate cements. Labella et al. (1994) found that by using MPS, the hardness, flexural strength, and diametral tensile strength of dental composites was significantly improved.

However, silane coupling agents also have been found to have different effects on the mechanical properties of other composites. Deb et al. (1996) and Nazhat et al. (1996) showed that MPS treated HA decreased the tensile strength and Young's modulus of the polyethylene composite. Since in both cases there was no chemical bonding between the silane and the polymer matrix, the decreased strength and modulus was explained by a plasticizing effect of the coupling agent. Surface treatment of β-crystalline meta-phosphate with the silane coupling agent MPS did not have an effect on the tensile strength of dental resin composites (Antonucci et al., 1991).

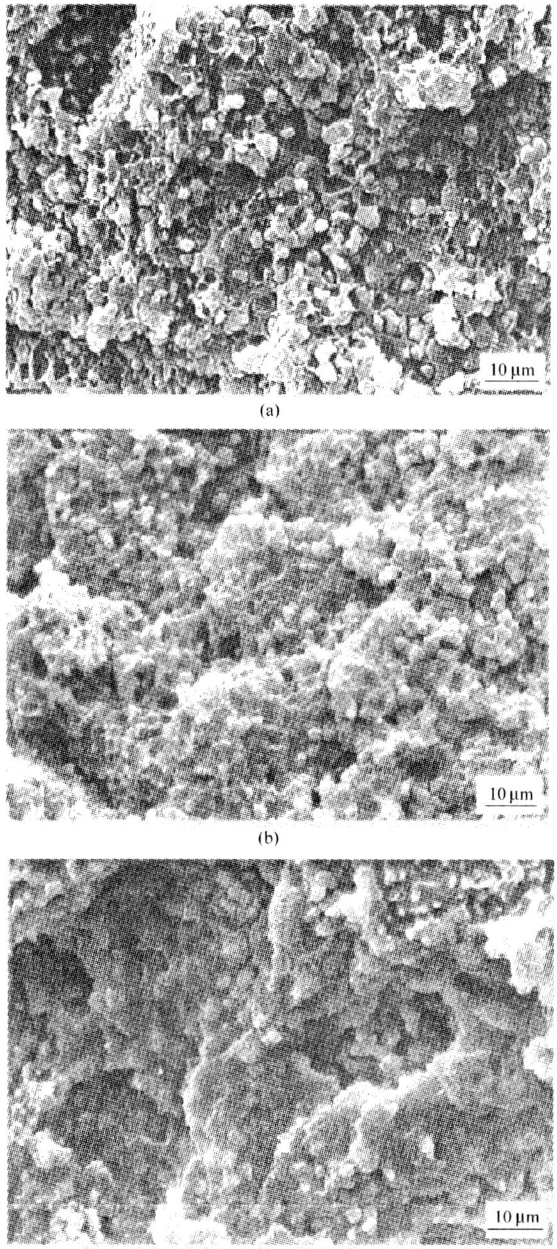

Figure 5. 9 Scanning electron microscopic pictures of the fracture surfaces of HA/Polyactive™ 30/70 composites. Unmodified HA particles (a) were loosely embedded in the polymer matrix, while HA particles modified with ethylene-maleic acid copolymer (b) or polyacrylic acid (c) showed more intimate contact with the polymer matrix (Liu et al., J. Mater. Sci. Mater. Med. 9: 23, 1998a)

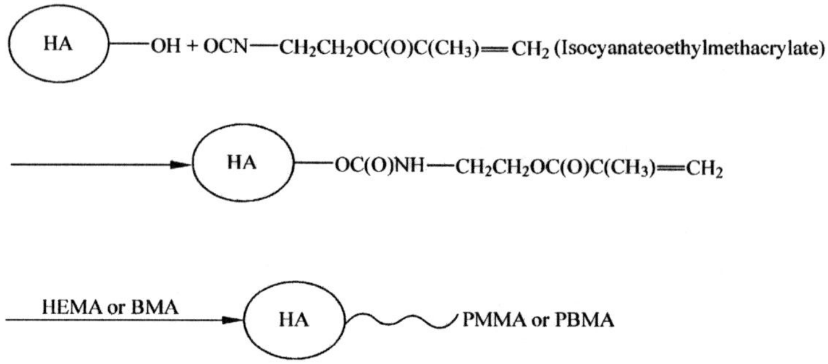

Scheme 1 PMMA or PBMA grafting using isocyanatomethacrylate as acouping agent.

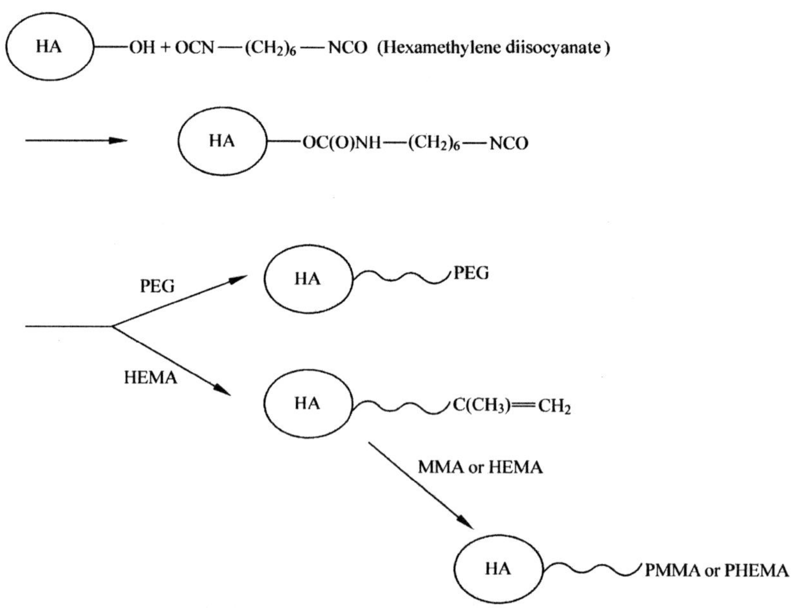

Scheme 2 Grafting PEG, PMMA and PHEMA to the surface of HA using hexamethylenediisocyanate and HEMA as coupling agents.

Figure 5. 10 Two schemes for grafting polymers to HA using ether isocyanatoethylmethacrylate or hexamethylenediisocyanate as coupling agents

In an effort to improve the adhesion of HA filler particles to a Polyactive™ matrix, Liu et al. developed two methods to introduce interactions between filler particles and the polymer matrix (Liu et al., 1996, 1997, 1998a – f). In the first method, ionic bonding and hydrogen bonding were introduced at the interfaces between the HA filler particles and Polyactive™ matrix by using polyacrylic acid (PAA) and ethylene-maleic acid (EMA) copolymers. It is well known that PAA and EMA can be firmly adsorbed onto the surface of hydroxyapatite. On the other hand, PAA and EMA can also form hydrogen bonds with polyethylene glycol which is present in the Polyactive™. Therefore, the adhesion of HA particles to the polymer matrix was enhanced by the presence of PAA or EMA at the interface. The mechanical properties of the resulting composites were significantly improved in both the dry and wet states. Scanning electron microscopy studies showed that HA filler particles modified with PAA or EMA adhered better to the polymer matrix (Fig. 5.9) (Liu et al., 1996, 1998a).

In the second method, isocyanate reagents were studied as coupling agents in nanoapatite/polymer composites. To improve the interface of the nanoapatite/polymer composites, Liu et al found that isocyanates were able to chemically bond to the surface of HA (Liu et al., 1998d, 1998e). Therefore, isocyanates and diisocyanates can be used to introduce chemical bonding between HA particles and various polymer matrices. Using this method, polyethylene glycol, poly (methyl methacrylate) (PMMA), poly (hydroxyethyl methacrylate) (PHEMA), and poly(butyl methacrylate) (PBMA) were covalently bonded to HA particles (Fig. 5.10) (Liu et al., 1998d, 1998f). Similarly, using hexamethylenediisocyanate as a coupling agent, covalent bonding between HA filler particles and the Polyactive™ matrix could be achieved, and the mechanical properties of the composites were greatly improved in the wet state (Liu et al., 1998b).

5.5 Tissue Engineering of Skin

The skin represents the interface between an organisms interior and its environment. It serves as a barrier to the environment outside of the body. As a barrier, the skin protects the body from physical trauma, electromagnetic radiation, loss of fluids, and bacterial invasion and possesses immune responsiveness. It has a surface area of 1.5 – 2 m^2 and is one of the largest organs of the body. The skin is composed of two layers in terms of its structure, namely, the epidermis and dermis (Fig. 5.11). The skin, like most of the tissues of the body, has the capability to repair its injuries. However, in some cases, self-healing is difficult or virtually impossible, as in the case of severe burns, venous ulcers, diabetic ulcers, and decubitus ulcers. It is estimated that every year there are about 12,000 patients who die from severe burns and thermal injuries, and about 3.5 million patients suffer from chronic wounds like venous ulcers, diabetic ulcers, and decubitus ulcers. In all these cases, skin regeneration poses a significant challenge to surgeons and researchers.

Tissue engineered skin research has primarily focused on the regeneration of the epidermis and /or repairing the dermis and has not focused on the regeneration of less important structures, like hair follicles and sebaceous glands (Teumer et al., 1998)

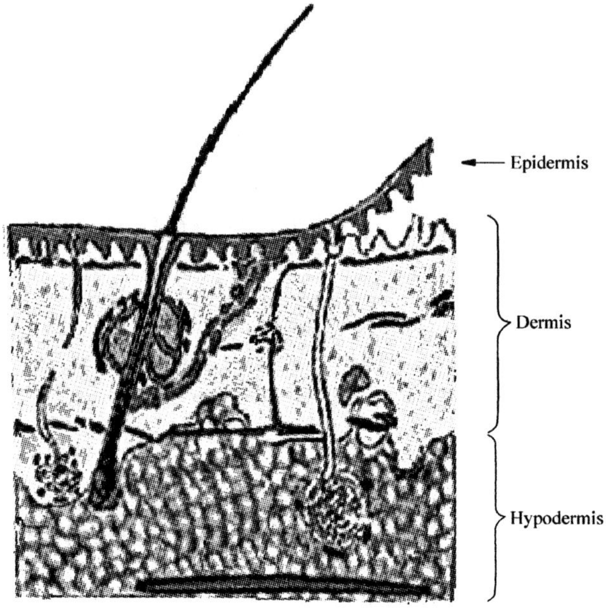

Figure 5. 11 Epidermis and dermis of skin

5.5.1 Regeneration of the Epidermis

Epithelization of a wound is the primary concern in wound closure, especially for patients with large wounds like severe burns. Without epithelial coverage as an outer barrier to the body, patients are more like to suffer from infections and bodily fluid loss.

Since the keratinocytes of the epidermis are primarily responsible for providing the barrier functions, some researchers have focused on their regeneration alone. In 1975, Rheinwald and Green established tissue culture techniques for the *in vitro* expansion of the human epidermis (Rheinwald and Green, 1975). In 1979, methods for the production of epithelia suitable for grafting were established using *in vitro* cell culture methods (Green et al., 1979; Phillips et al., 1989). Epidermal keratinocytes obtained from a small patient biopsy were used and cultured *in vitro*. Using this method, enough thin, multilayered epidermal sheets were generated to resurface the body of a severely burned patient (Gallico et al., 1984). It has been demonstrated that once transplanted, the epidermal sheets quickly form an epidermis and reestablish epidermal coverage in patients (Compton, 1993). With time, the cultured epithelial autograft will lead to the formation of new connective tissue or the "neodermis" immediately underlying the epidermis. However, the clinical use of this cultured epidermis often is unsatisfactory. Blistering, scaring, and wound contraction are significant problems (Scheridan and Tompkins, 1995).

A gene modification method was used to modify the cultured epithelium to enhance the function of the epithelium. Eming et al. (1996) developed a genetically modified skin graft which would function as a cell-based vehicle for the local synthesis and delivery of wound-healing growth factors. Using retroviral-mediated gene transfer, they introduced stable copies of the genes encoding platelet-derived growth factors (PDG F-A) and the insulin-like growth factor-1 (IGF-1) into cultured human diploid keratinocytes. After stable integration of these genes, the cells secreted significant levels of these growth factors of PDG F-A and IGF-1, respectively. The modified cells were grown to confluence, detached as multicell-layered epithelial sheets, and transplanted to athymic mice. After seven days, the grafts secreting PDG F-A or IGF-1 differentiated into a stratified epithelium comparable to unmodified cells. Most importantly, the newly synthesized connective tissue layer subjacent to the PDG F-A-modified grafts was significantly thicker and showed an increase in cellularity, vascularity, type I collagen, and fibronectin deposition when compared to control grafts of unmodified cells or grafts expressing IGF-1.

5.5.2 Regeneration of the Dermis

The regeneration of the dermis is of equal importance to the regeneration of the epidermis. Dermal repair after severe wounds is important to restore the skin's elasticity, flexibility, and strength. Scar tissue lacks these characteristics and is permanent because the dermis has very little capacity to regenerate.

It has been found that the use of the wound cover will provide a structural framework for wound healing. If this structure is not present, fibroblastic tissue will lay down collagen in a random way, resulting in a disordered dermis (scar). By providing scaffolding for the new vascular and mensenchymal tissue, an orderly and controlled restructuring of the dermis can take place without scar formation.

Yannas et al. developed one of the earliest tissue engineering approaches to improve dermal healing (1980a, 1980b, 1982, 1990). Yannas and co-workers developed an artificial skin based on reconstituted bovine collagen and glycosaminoglycans (GAG) (predominantly chondroitin-6-sulphate). The addition of GAG improved the mechanical properties of the reconstituted collagen matrix, also allowed better control of pore structures, and yield improved resistance to collagenase degradation (Dagalaskis et al., 1980; Yannas, 1990). By varying the cross-link density, the collagen/GAG content and the proportion of highly crystalline collagen, it was possible to modulate the biological activity and mechanical behavior, the pore structure and biodegradation rate of the skin templates (Yannas, 1990). The hydrated gels of covalently cross-linked graft copolymers appeared to possess remarkable biological activity. When used in skin wound models in guinea pigs, wound contraction was significantly reduced, and the grafts, seeded with keratinocytes, successfully regenerated skin with a dermis and epidermis clearly distinct from scar tissue (Yannas, 1989, 1990). Later, Boyce et al. (1988, 1989) developed a bilayered laminated structure based on similar collagen – GAG copolymers to prevent the seeded keratinocytes from growing into the porous membranes, while improving the seeding efficiency of the keratinocytes on the membranes (Hansbrough et al., 1989). In an athymic mice study, human fibroblast and keratinocyte seeded skin substitutes led to complete wound closure of full-thickness skin defects in 2 to 4 weeks. The healed wounds developed a fully keratinized desquamating epidermal surface. They also found pigmentation of grafted epidermis and nerve ingrowth in the neodermis. A commercial version of this collagen – GAG material (Integra™, manufactured by Integra Life Science) has been approved for use under split-thickness autografts in burns (Burke et al., 1981; Heimbach et al., 1988).

Synthetic biodegradable polymers were also used as scaffolding materials to grow human fibroblast as dermal grafts. Polyglycolic acid (PGA) and a polyglacitin 910 (PGL) mesh containing confluent cultured human fibroblasts were applied to full-thickness wounds in athymic mice studies (Hansbrough et al., 1992a). On top of this mesh, expanded human meshed skin grafts were placed. It was found that in a 99-day period, the PGA/PGL – fibroblast grafts were incorporated into the wound, and epithelialization from the skin bridges proceeded rapidly across the surface of the grafts. They also observed minimal inflammatory reactions of the grafts and basement membrane formation at the dermal – epidermal junction of the epithelialized interstice.

A similar construct using PGA mesh (Vicryl™, Johnson and Johnson) and cultured high density dermal fibroblasts (Dermagraft, Advanced Tissue Sciences, La Jolla, CA) was used in clinical studies and showed promising results (Hansbrough et al., 1992b). This construct is also used for the treatment of diabetic foot ulcers.

Recently, Van Dorp et al. (1998) used a biodegradable copolymer of poly (ethylene/poly (butylene terephathalate) (Polyactive™) as a substrate to construct a dermal substrate. They cultured fibroblasts on the porous bilayered Polyactive™ substrate for 3 weeks to obtain a fibroblast-populated skin graft. During the 3-week incubation period, the pores of the substrates were filled with fibroblasts and extracellular matrices (collagen type Ⅰ, Ⅲ, Ⅳ and laminin) (van Dorp et al., 1999). In an animal study using pigs, they found that rapid vascularization occurred within 4 days after transplantation. The newly generated dermal architecture resembled that of healthy skin and thus they concluded that the treatment of full-thickness skin defects with this fibroblast-populated Polyactive™ substrate leads to satisfactory dermal regeneration.

5.5.3 Living Skin Equivalent

Bell and his colleagues have developed a "living skin equivalent" made from three-dimensional collagen lattices seeded with autologous fibroblasts (Bell et al., 1981; Coulomb et al., 1986a, 1986b). In this configuration, fibroblasts stop dividing, become synthetically active, and start remodeling their extracellular environment. Once contraction by the fibroblasts is stabilized, epidermal keratinocytes can be successfully placed onto the surface of the dermal equivalent, yielding a living skin equivalent. Experimental animal results showed that the dermal layer was remodeled over the next 10 weeks after the graft is placed. Epithelial growth could be seen at the edges of the wound (Bell et al., 1991; Nusgens et al., 1984). A commercial version of this "living skin equivalent" is called Apligraf™ (Organogenesis Inc., Cambridge, MA).

Another skin substitute, composed of both epidermal and dermal elements, is being studied (Zacchi et al., 1998). One of the scaffold biomaterials used was derived from the total esterification of hyaluronan with benzyl ester (HYAFF-11™, Fidia Advanced Biopolymers, FAB, Abano T., Italy). The molecule is synthesized starting from 80 – 200 kDa sodium hyaluronate. The scaffolds were constructed as either (1) a nonwoven mesh made of 20 mm thick fibers or (2) 20 mm thick membranes with laser-made holes of 40 mm in diameter (LaserSkin™, Fidia Advanced Biopolymers, FAB, Abano T., Italy). Keratinocytes and fibroblasts were grown separately for 15 days within two different types of biomaterials. Cells then were cocultured for an additional period of 15 days, after which samples were taken and processed with either classic or immunohistochemical stains. Results showed that (1) human fibroblasts and keratinocytes can be cultured on hyaluronic acid-derived biomaterials and that (2) the pattern of expression of particular dermal – epidermal molecules is similar to that found in normal skin. This skin equivalent has the potential to be used in the treatment of both burns and chronic wounds.

5.6 Tissue Engineering of Bone

Bone, together with cartilage, makes up the human skeletal system. Bone is a highly vascular, constantly changing, specialized type of mineralized connective tissue which serves the following three functions:

(1) Bone provides mechanical support and is the site of muscle attachment for movement;

(2) Bone protects the viscera and surrounds bone marrow;

(3) Bone serves a metabolic function, storing calcium, phosphorus, and other ions.

Bone consists of specialized bone-resorbing and bone-forming cells and an extracellular matrix. The matrix is composed of type I collagen fibers, noncollagenous proteins, and mineral crystals of calcium phosphate (hydroxyapatite).

Collagen fibers and crystals of bone salts are arranged in flattened plates (lamellae) that are variously oriented so as to give bone its strength.

There are two types of bone in terms of their structures. They are cortical bone and trabecular bone (Fig. 5.12).

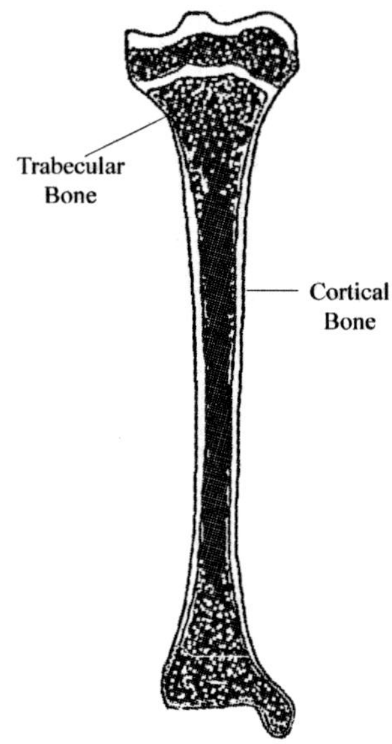

Figure 5.12 Cortical bone and trabecular bone

Cortical bone occurs in the walls of the shafts of long bones and as the outer shell of flat bones such as the pelvis. Cortical bone is dense with respect to trabecular bone and contains closely spaced groups of concentric lamellar bone rings called osteons. Cortical bone functions mechanically in tension, compression, and torsion. Trabecular bone occurs in the interior of flat bones and near the ends of long bones in the region of the joint surface. It consists of a porous array of interconnected lamellar bone rods and plates (trabeculae). Trabecular bone mainly functions in compression. The spaces between the trabeculae contain the bone marrow. Bone is a dynamic, highly vascularized tissue with the ability to heal and to remodel.

The normal fracture healing of bone only requires temporary immobilization of the fractured area. And the healing process results in the reformation of organized, unscarred bone tissue. However, in some cases, bone tissue does not heal completely (nonunion) due to fracture comminution or bone loss, infection, loss of blood supply, disease, or inadequate fracture management (Yaszemski et al., 1995). In these cases, certain therapies are needed to restore the bone structure and/or function.

Bone substitutes are often required to replace damaged bone tissue due to disease, trauma, or surgery. Conventional methods for the reconstruction of surgical osseous defects is dependent on an adequate supply of autogenous (host) or allogenic (donor) bone. A bone autograft is widely considered to be the best implant for repairing bone defects. However, the amount of autogenous bone available for transplantation is limited, particularly in children. Also, the harvesting operation carries the risk of postoperative complications.

Allogenic bone has been successfully used in osseous reconstruction and offers several advantages over autogenous bone, including the avoidance of a harvesting operation, ease of manipulation, and potentially unlimited material in the bank form. Nevertheless, the possibility of transmitting diseases from the donor to the recipient raises doubts about its future. Long times are needed for the resorption and the replacement of allografts by new bone, and the antigenic activity of banked bone are serious disadvantages when compared to autografts.

Synthetic biomaterials (metals, ceramics, and polymers) offer many advantages over autografts and allografts for the treatment of osseous defects in many aspects. Synthetic biomaterials are available in large quantities. There are no potential disease transmission problems for synthetic biomaterials. The mechanical properties of the synthetic biomaterials can be tailored to meet the specific needs of the patient. Drugs and bioactive agents can be incorporated into the biomaterials for specific purposes. However, the use of synthetic materials has limitations, too. For example, metals have high mechanical stiffness and may result in stress shielding and bone resorption due to mismatch in the elastic modulus. Ceramics are usually too brittle to be used as load-bearing materials for bone replacement. Polymers have poor mechanical properties as compared to those of cortical bone.

Clearly, there is a need for an alternative bone graft that, ideally, possesses the properties of living bone. With the recent developments in the field of bone tissue engineering, these tissue engineered implants are likely to become available for clinical use in the coming years.

There are two commonly used techniques in bone tissue engineering.

5.6.1 Bone Growth Induction

Biomaterials used for bone tissue engineering basically can be categorized as osteoconductive and osteoinductive materials.

Osteoconductive biomaterial scaffolds can be used to guide bone tissue growth. Of the biomaterials, calcium phosphate [hydroxyapatite (HA), tricalcium phosphate (TCP), octacalcium phosphate (OCP)] and Bioglass are the two types of ceramic materials which have drawn great attention.

Porous HA has the ability to induce bone formation in both a bony site and a soft tissue site (Klein et al., 1994, Ripamonti, 1991) It is considered to be osteoconductive and is frequently used as filling a materials to fill in the small bone defects to induce bone tissue growth.

Since the major disadvantage is that these ceramic materials are brittle and have insufficient strength for use in load-bearing situations, they are often used as coatings on metal implants or used together with polymers as composites.

After being filled with sufficient amounts of HA, a non-osteoconductive polymer will be turned into an osteoconductive composite. Polyethylene, polybutyrate, PLLA, PGLA Polyactive™, and PPF are the polymers frequently used in such composites (Higashi et al., 1986; Liu et al., 1997, 1998b; Luklinska and Bonfield, 1997; Thomson et al., 1999; Verheyen et al., 1992b; Yaszemski et al., 1996b).

Among these polymeric biomaterials, biodegradable polymers are of great interest.

These ceramic/polymer composites can be premanufactured *in vitro*; they can be also shaped *in vivo*. An example is the injectable PPF/TCP biodegradable polymeric bone cement developed to replace the widely used nonbiodegradable PMMA bone cement. PPF bone cement can be injected into bone defect without surgery. TCP will act as an osteoconductive component to promote bone tissue ingrowth and therefore restore normal bone geometry and/or stabilize a bone fracture (Peter et al., 1998a; Yaszemski et al., 1995, 1996a).

To make scaffolds more osteoinductive, researchers also have incorporated bone growth factors such as bone morphogenetic proteins (BMPs) and TGF-β in the scaffolds. Bone morphogenetic proteins (BMPs) are a subclass of the transformed growth factor-beta (TGF-β) superfamily that provides cues for bone development and regeneration. BMPs and TGF-β have different effects on bone formation. Several BMPs are able to induce bone ectopically (subcutaneously or intramuscularly). TGF-β does not do so but can promote bone formation in the vicinity of bone (Nimni, 1997).

It has been found that when enough BMPs were loaded into an acid-demineralized bone matrix, the matrix induced new bone formation (Rapamonti et al., 1992; Reddi, 1998a; Yasko et al., 1992). Collagen carriers have been also used to deliver BMPs to regenerate bone in a variety of animal studies such as craniotomies (Marden et al., 1994), mandibular reconstruction (Toriumi et al., 1991), and large segment bone defects (Cook et al., 1994).

Biodegradable polymers such as poly(α-hydroxy acids) and PLGA are also chosen as carriers for growth factors (Hollinger and Leong, 1996; Puleo et al., 1998).

Growth factors also may be applied to orthopedic metalic implants to promote bone remodeling around implants. Lind et al. (1996a, 1996b) showed that by simply applying TGF – β to the surface of TCP, HA, or TCP coated implants through adsorption, bone formation in the periprosthetic gap was improved. Since the growth factor is only physically adsorbed on the surface of the implant, there is little control over the delivery, including the release/retention and orientation, of growth factors. Covalently attaching biomolecules to the surface of implants is an alternative way to promote bone remodeling around the implant. This method has been used to immobilize peptides, enzymes, and adhesive proteins on different metallic biomaterials (Endo, 1995; Nanci et al., 1998; Peleo, 1995, 1997; Ferris et al., 1999). For example, a peptide sequence, arginine – glycine – aspartic acid (RGD), was immobilized onto gold coated titanium rods through a gold-thiol reaction. Then the RGD coated titanium rods were implanted in rat femoral canals. A histomorphometric analysis of cross-sections perpendicular to the implant long axis showed a significantly thicker shell of new bone formed around RGD coated implants at 2 weeks. A significant increase in bone thickness for RGD-modified implants was also observed at 4 weeks, while the bone surrounding the controls did not change significantly in thickness (Ferris et al., 1999).

5.6.2 Cell Delivery

This technique uses autologous osteoblast or osteoprogenic cells. The harvested osteoblast and osteoprogenic cells are cultured *in vitro* to expand the population of the cells in order to get enough cells for the transplant. Then the cultured cells are seeded in biomaterial scaffolds and transplanted. After transplantation, the cells may start to proliferate, secrete a matrix, and release growth factors as the scaffold vascularizes and slowly degrades.

This technique provides a way to use a patient's own autologous osteogenic cells to treat bone defects. Therefore, it eliminates the complications of immune rejection, donor site morbidity, and limited availability.

Several cell sources have been identified as osteogenic cells. These cells are embryonic cells (Gearhart, 1998; Thomson et al., 1998), mesenchymal stem cells (Ohgushi et al., 1993; Ohgushi and Caplan, 1999), osteoprogenitor cells from bone marrow (de Bruijn et al., 1996, 1999,), and periosteal cells (Vacanti et al., 1995). Bone marrow is suitable for an osteogenic source because it contains subpopulations of osteoprogenitor cells. It has been shown that bone marrow cells can maintain their viability and differentiate into osteogenic cells. Such cells can maintain their osteogenic capacity when implanted subcutaneously by showing new bone formation (de Bruijn et al., 1996, 1999; Ishaug-Riley et al., 1997).

Biodegradable polymers are extremely attractive as scaffolding materials because the scaffolds can degrade gradually over time and finally disappear. Therefore, no alien materials will remain in the body as the bone defects finally heal.

Poly(α-sshydroxy acids) are promising substrates for cell transplantation. Among them, poly(lactic acid) (PLA), poly(glycolic acid) (PGA) and poly (lactic-co-glycolic acid) (PLGA) are currently under investigation. It is important to note that some of the formulations of poly(α-hydroxy acids) are approved by the U. S. Food and Drug Administrations for use in certain surgical procedures.

Animal experiments showed very promising results using periosteal cell/ polymer constructions. PGA scaffolds were seeded with periosteal cells; then the construct was placed into either rats' or rabbits' nonunion defects. Early bone formation was found in both cases (Breitbart et al., 1998; Vacanti et al., 1995).

Polyphosphazenes, poly (anhydride-co-imides) and poly (propylene fumarate) also have been investigated as scaffolds for bone cell delivery (Attawia et al., 1999; Laurencin et al., 1996a, 1996b, 1998a; Peter et al., 1998a).

When the scaffolds are porous, the mechanical properties or the compressive strength is quite weak. Therefore, certain reinforcement techniques were employed to strengthen the polymer scaffolds. HA fibers and TCP particles were frequently used as fillers in making the scaffolds (Devin et al., 1996; Peter et al., 1998a; Thomson et al., 1995).

Porous bioactive ceramics are also suitable substrates for the attachment of cells. They also were used as scaffolds for cell delivery. They can be manufactured into three-dimensional scaffolds to support bone growth both *in vitro* and *in vivo* (Livingston et al., 1999; Oonishi et al., 2000). In a study, porous surface modified bioactive glass-ceramics (pSMC) were investigated as tissue engineering scaffolds in the long bone defect of rats. Rat bone marrow stromal cells were either seeded on pSMC 2 hours prior to implantation (primary), or cells were expanded on pSMC for 2 weeks to synthesize the bone prior to implantation (hybrid). Defects were treated randomly with pSMC, primary, hybrid, or left untreated (sham) to compare healing rates at 2, 4, and 12 weeks. At 2 weeks, long bones treated with the hybrid and primary constructs had 40% more bone in the defect than pSMC and the hybrid had the highest stiffness, which was comparable to intact bone. Primary and hybrid had stiffness and strength comparable with intact bone by 4 weeks. At 12 weeks, there was a 40% bony ingrowth and 40% reduction in percent scaffold for all treatment groups. PSMC achieved similar mechanical properties to those of intact bone by 12 weeks. Both tissue engineered constructs achieved early and similar rates of repair (Livingston et al., 1999).

In some other studies, fresh marrow cells or cultured mesenchymal stem cells were combined with porous ceramics and implanted into rat (Kadiyala et al., 1997; Ohgushi et al., 1989) or canine segmental bone defects which cannot heal by themselves (Bruder et al., 1998). Two months following implantation in orthotopic sites, the defects that received ceramics without marrow showed only a minimum amount of new bone, while the marrow/porous ceramic construct showed bony continuity brought about by regenerated bone tissue in and around the ceramics, which indicates the healing potential of the construct.

In most of the cases, the cultured cells are incubated for only a few hours in the porous biomaterial scaffold prior to implantation. Recently, an alternative way was developed to create a true tissue engineered autologous bone graft (de Bruijn et al., 1999). In this approach, the (pre) osteogenic bone marrow stromal cells are cultured, expanded, and subsequently seeded on a biomaterial scaffold. The cells are cultured on the scaffold for a sufficient time to induce osteogenic differentiation and extracellular bone matrix production. The advantages of this approach are twofold. First, the cultured cells have already started to produce a bone matrix *in vitro* and will continue this process *in vivo*, resulting in an accelerated bone healing process after implantation. Second, the cultured bone matrix will contain various proteins and bone growth factors that are expected to induce bone formation *in vivo* as described above. Therefore, by culturing autologous bone marrow stromal osteoprogenitor cells *in vitro* on a suitable scaffold, it is possible to produce a true autologous bone graft equivalent.

5.7 Tissue Engineering of the Peripheral Nervous System

Nerve injury may result from mechanical, thermal, chemical, or pathological damage. The current clinical methods for treatment of nerve damage are surgical end-to-end anastomosis and autologous nerve grafts. Surgical end-to-end repair is the direct reconnection of individual nerve fascicles or bundles of fascicles. However, when the two ends of the damaged nerve stumps are not adjacent to each other, the nerve ends cannot be reconnected. In this case, an autologous nerve graft can be used to repair the peripheral nerve over a gap. However, with this method, healthy nerves have to be sacrificed and that frequently leads to multiple surgical procedures. Other disadvantages include the loss of function at the donor site, the mismatch of nerve cable dimensions between the donor graft and the recipient nerve, etc. A promising alternative is to use nerve guidance channels.

The mechanism of using a nerve guide for the regeneration of a damaged peripheral nerve is, when using a nerve guide, that a gap is left between the nerve stumps sutured into either end of the tube. A matrix composed predominantly of fibrin spontaneously forms in the gap from exudate derived from the stumps. The regenerating nerve fibers are allowed to grow toward the distal nerve stump in this matrix, while neuroma formation and ingrowth of fibrous tissue into the nerve gap is prevented (Fig. 5.13). Furthermore, inside the nerve guide, tropic and trophic factors, produced by the distal nerve stump, can accumulate. Trophic factors are necessary for the survival and growth of the damaged axons, and tropic factors are necessary for the growth direction of the regenerating axons. The growth direction of the regenerating axons is not only caused by a mechanical effect (the wall of the nerve guide), but also by a chemical effect (the gradient of trophic factors).

Most of the NGCs have dense walls which serve the purposes of both preventing scar tissue from invading and keeping the tropic and trophic factors produced by the distal nerve stump. However, one concern regarding the tubulation of nerves in the impermeable condu its is the limitation of a nutrient supply to the regenerating tissue inside the lumen of the NGCs. Recently, several authors have addressed this problem and used NGCs with permeable wall

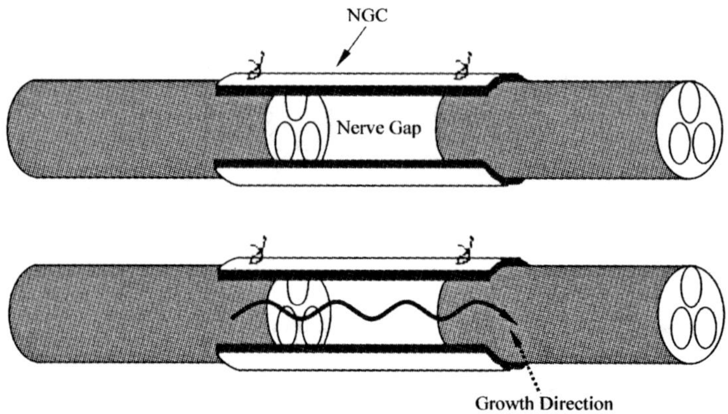

Figure 5.13 Use of a NGC to guide peripheral nerve regeneration

structures (den Dunnen et al., 1993a,b; Widmer et al., 1998b). A NGC with an optimum porous structure will allow the influx of nutrients and growth factors from the external environment and help concentrate factors released by the proximal nerve stump.

Besides the morphology, the diameter of the NGC is another important factor. A NGC in which the nerve fits too loosely will fail to support the regenerating nerve, while on the other hand, a NGC in which the nerve fits too tightly will cause damage to the nerve. It is generally considered that a cross-sectional area of 2.5 to 3 times the area of the nerve is the ideal opening diameter of the NGC (Widmer et al., 1998b).

The material used in making the NGC played a very important role in this technique. There are some general requirements for the NGC material: first the NGC material should be biocompatible. Second, the NGC material should be pliable yet strong enough to maintain the shape of the NGC over the regeneration period. A biodegradable NGC is preferred.

5.7.1 Materials

In the past, several materials, either of biological origin (Cataltepe et al., 1993, Colin and Donoff, 1984; Walton et al., 1989) or synthetically fabricated (both biodurable and biodegradable) (den Dunnen et al., 2000; Molander et al., 1982), have been tried for this purpose.

5.7.1.1 Materials of Biological Origin

Various materials of biological origin such as laminin, fibronectin, and collagen have been used for NGCs (Furnish and Schmidt, 1998). NGCs derived from laminin, fibronectin and collagen all showed improved regeneration over a 4 mm or 10 mm rat sciatic nerve gap compared to silicone NGC controls. In addition, the vein graft has also been used as a NGC (Walton et al., 1989). To improve the regeneration effect of the vein, a collagen coating can be applied to the lumen of the vein grafts, or the vein can be turned inside out to expose the collagen – rice layer to the lumen (Wang et al., 1993a, 1993b). Both NGCs showed enhanced regeneration across a 10 mm defect in the rat sciatic nerve compared to standard vein grafts. Other materials like synovial (Lundborg and Hansson, 1980) and perineurial (Restrepo et al., 1983) were also explored for use in NGCs.

The problems associated with materials of biological origin are immunogencity and poor mechanical properties.

5.7.1.2 Synthetic Materials

1. Nonbiodegradable Materials

Non-biodegradable polymer materials have been used in making NGCs. Poly(acryl-co-vinyl chloride) (Aebischer et al., 1990), polycarbonate (Harvey et al., 1994), polyethylene (Cordeiro et al., 1989), polysulfone (Aebischer et al., 1989a), and silicone rubber were used as NGCs. Since the electrical charge of the materials will affect nerve regeneration, the piezoelectric material, polyvinylidene fluoride (PVDF), and a charge-bearing material, polytetrafluoroethylene (PTFE), were used to construct NGCs. It has been shown that neurite extension was enhanced both *in vitro* and *in vivo* on these two materials (Aebischer et al., 1987; Valentini et al., 1992). Recently, an electrically conducting polymer polypyrrole was used to construct NGCs (Schmidt et al., 1997). In an *in vitro* study, electrical stimulation applied to the polypyrrole film significantly enhanced neurite outgrowth. Polypyrrole conduits also showed the ability to guide nerve regeneration across 10 mm defects in rats.

However, NGCs fabricated with these materials remain *in vivo* as foreign bodies and may cause foreign body reactions. For example, silicone nerve guides were used successfully in nerve regeneration to reconstruct peripheral nerves in three patients. However, after 2 years, the patients began complaining about secondary nerve impairment and irritation at the implantation site (Merle et al., 1989). A second operation was needed to remove the biodurable silicone rubber nerve guides. Because the removal of a biodurable nerve guide might damage the regenerated nerves, nerve guides should degrade immediately after serving their function.

2. Biodegradable Materials

Synthetic biodegradable polymer materials such as polyesters, polyurethanes, and polyphosphazenes have been used in making NGCs. In an ideal situation, the degradation rate of the nerve guide should be in accordance with axonal growth rates. Besides, biodegradable nerve guides have to fulfill several other requirements. The biomaterial should be biocompatible, which means that it should be noncytotoxic, noncarcinogenic, nonimmunogenic, nonmutagenic, and cause no irritation or allergic response, either local or systemic. Furthermore, a nerve guide should be flexible, (semi) permeable, and easy to apply in microsurgery. Preferably, a nerve guide should also be transparent. Besides the transparency, which allows accurate observation of the nerve stumps when telescoping them into the nerve guides, (semi) permeability of the wall of the nerve guides has a positive effect on nerve regeneration.

One of the first synthetic biodegradable polymers used in NGCs was poly (lactic-co-glycolic acid) (PLGA). Suture mesh made from PLGA was used to bridge 7 – 9 mm defects in the tibial nerves of rabbits. Although the regenerated nerve cable had a slightly different morphology from normal nerves, these studies demonstrated that PLGA is a low irritant to the regenerating nerve and that synthetic materials can be used as NGCs (Molander et al., 1982). Later, a semicrystalline copolymer of L-lactic acid and ε-caprolactone (50:50) was prepared and tested (Grijpma et al., 1991). This material proved to be noncytotoxic (den Dunnen et al., 1993a), and the nerve regeneration after reconstruction with this type of nerve guide was good. However, after 2 years, fragments of the nerve guide material were still present around the regenerated nerve, causing a chronic foreign body reaction with scar tissue formation (den Dunnen et al., 1993b). To overcome this problem, D-lactide was added to obtain a biomaterial with a lower crystallinity than PLLA and consequently a higher degradation rate (den Dunnen et al., 2000). The first results with this poly(96L/4D-lactide) are promising (den Dunnen et al., 2000).

Biodegradable polyurethane was also used to construct a NGC, and this nerve guide channel supported regeneration across an 8 mm gap in the rat's sciatic nerve (Robinson et al., 1991). This nerve guide, however, had several disadvantages; it was relatively hard and brittle, and even more serious, the degradation products of the polyurethane were cytotoxic. Further-more, the polyurethane did not degrade completely.

Polyphosphazene was also tested as a NGC and showed promising results (Langone et al., 1995).

5.7.2 Methods for Promoting Nerve Regeneration Inside NGCs

There are several methods for promoting nerve regeneration inside NGCs. These methods are

(1) Use of specific biochemical factors;
(2) Incorporation of beneficial matrix proteins;
(3) Addition of neuronal support cells.

5.7.2.1 Use of Specific Biochemical Factors

Neurotropic and neurotrophic factors can stimulate nerve regeneration. Therefore, these factors can be delivered locally by incorporation in NGCs. Factors which have been identified to be beneficial to nerve regeneration are nerve growth factors (NGF), brain derived neurotrophic factors (BDNF), insulin-like growth factors (IGF-1 and IGF-2), transferrin, platelet-derived growth factors (PDGF-BB and PDGF-AB), basic fibroblast growth factors (bFGF), acidic fibroblast growth factors (aFGF), ciliary neurotrophic factors (CNTF), and interleukin-1 (IL-1) (Furnish and Schmidt, 1998).

An example of using a biochemical factor is incorporating bFGF into a NGC made from ethylene vinyl acetate. A 15 mm gap in the rat sciatic nerve was bridged using this bFGF impregnated NGC (Aebischer et al., 1989).

5.7.2.2 Incorporation of Beneficial Matrix Proteins

A lot of matrix proteins can be used in NGCs. Fibrin, collagen, muscle, and hyaluronic acid were used to promote nerve regeneration inside NGCs.

Silicon tubes encasing a collagen – glycosaminoglycan matrix yielded improved regeneration across a 10 mm defect in a rat sciatic nerve compared to that of saline-filled tubes (Chang et al., 1990). Silicon tubes filled with fibrin precursors from plasma enhanced nerve regeneration over a gap of 15 mm (Williams et al., 1987). Growth factors can also be used together with matrix proteins. In an experiment, polyethylene tubes filled with a gel of collagen and aFGF showed greater numbers of myelinated axons compared to collagen-filled tubes (Cordeiro et al., 1989).

Recently, muscle tissue was found to be able to speed up the recovery of nerve damage. Meek et al. used denatured muscle tissue to fill a biodegradable p (DLLA-E-C L) nerve guide. The filled nerve guide was used to bridge a 15 mm gap. The muscle filled nerve guide showed faster functional nerve recovery compared to a nerve reconstruction of a 10 mm gap with the same P(DL-lactide-ε-caprolactone) nerve guide filled with PBS saline (Meek et al., 1996). In a clinical study, Battiston et al. used fresh skeleton muscle tissue to fill in veins (to prevent vein collapse and support axon regeneration) to reconstruct nerve defects ranging from 5 – 60 mm in length. The clinical results seem to suggest that this type of muscle – vein combined graft is better than other kinds of artificial and biological conduits in terms of clinical results (Battiston et al., 2000).

5.7.2.3 Addition of Neuronal Support Cells

Neuronal support cells can be cultured on the internal lumens of NGCs. The purposes of seeding neuronal support cells on NGCs are twofold. First, the seeded cells can produce various neurotropic and neurotrophic factors which will promote nerve regeneration. Second, the seeded cells provide a good substrate for axons to elongate. Schwann cells have been used as neuronal support cells because they can secret neurotrophic factors (CNTF, BDNF, NGF, and IGF-1). In an animal experiment, Schwann cells were seeded on the internal lumen of poly-L-lysine coated polyethylene tubes and implanted into rats to bridge a 10 mm gap. Two weeks following implantation, a significantly large amount of total axons and myelinated axons were found in the Schwann cell seeded nerve guides (Bryan et al., 1992). Levi et al also used human Schwann cell seeded PAN/PVC conduits. This cell seeded nerve guide increased axonal regeneration over an 8 mm gap in the rat sciatic nerve (Levi et al., 1994).

In summary, nerve guidance channels are synthetic or natural conduits that help direct axons (nerve protrusions that guide regeneration) sprouting off the regenerating nerve end. They provide a conduit for the diffusion of growth-enhancing molecules secreted by the damaged nerve stumps and minimize infiltrating scar tissue that may impede regeneration. Furthermore, nerve guidance channels can be fashioned to actively stimulate the nerve to regenerate, for example, by the release of specific biochemical factors or by the incorporation of beneficial matrix proteins or electrical stimuli. It has to be mentioned that electrical stimuli may provide a potential means to promote nerve regeneration. Some *in vitro* studies have shown that when using polypyrrole as a cell culture substrate, neurite outgrowth was significantly enhanced by passing an electrical current through the polypyrrole substrate films (Schmidt et al., 1997). Although most research in this area has been applied to the development of alternatives to peripheral nerve grafts, studies are beginning to focus on creating guidance channels to promote regeneration of axons in the spinal cord.

5.8 Tissue Engineering of the Heart Valve

Currently, heart valve substitutes are of two principal types: mechanical prosthetic valves with components manufactured of nonbiological material (e.g., polymer, metal, carbon) or tissue valves which are constructed, at least in part, of either human or animal tissue. It is estimated that at least 60,000 substitute valves are implanted in the United States and 170,000 worldwide each year (Schoen and Levy, 1999). There are some serious problems associated with the current heart valve substitutes, both mechanical and tissue valves. The problems are (1) thromboembolism, thrombosis, and anticoagulation-related hemorrhage; (2) prosthetic valve endocarditis (infection); (3) structural dysfunction (i.e. failure or degeneration of the prosthetic biomaterials; and (4) nonstructural dysfunction [i.e. a diverse array of complications which includes tissue overgrowth, paravalvular leak, hemolysis, and other extrinsic interactions of host tissues with a valve (Schoen and Levy, 1999)]. An even more serious problem for both tissue and mechanical valves and valved conduits is that they are unable to grow, be repaired, or be remodeled.

Tissue engineering provides an exciting concept for growing an ideal heart valve substitute. The ideal heart valve substitute is a nonobstructive, nonthrombogenic tissue valve substitute which will last the lifetime of the patient, provide ongoing remodeling and repair of cumulative injury, and potentially grow in maturing recipients.

Though still in their early stage, tissue engineered heart valves showed very promising results in animal experiments.

The tissue engineering approach again includes the use of biodegradable scaffold and cells. First, an anatomically appropriate biodegradable polymer scaffold is manufactured. Then the cells (fibroblasts and endothelial cells) are seeded in the polymer scaffold. After culturing *in vitro* for a certain period of time, the cell seeded construct is implanted. The transplanted autologous cells will generate a proper matrix on the polymer scaffold in a physiological environment. The polymer scaffold will gradually degrade and finally disappear and leave a regenerated heart valve.

Several biodegradable polymers were used as scaffold materials in heart valve tissue engineering. Synthetic polymers: polyglycolic acid (Shinoka et al., 1995, 1996; Sodian et al., 2000), polyhydroxyalkanoate (PHA) (Sodian et al., 2000), and poly-4-hydroxybutyrate (P4HB) (Sodian et al., 2000); natural polymers: fibrin gel (Ye et al., 2000) and tissue acellular matrix (Bader et al., 1998).

The cells used to construct the heart valve are fibroblasts, endothelial cells, and smooth muscle cells. They are usually isolated from an explanted artery or vein.

The first tissue engineered heart valve experiment was carried out using mixed cell populations of endothelial cells and fibroblasts which were isolated from explanted ovine arteries. A synthetic biodegradable scaffold constructed from polyglycolic acid fibers was seeded with fibroblasts *in vitro* and grew to form a tissue-like sheet. This tissue was subsequently seeded with endothelial cells, which formed a cellular monolayer coating around the leaflet. Using these constructs, autologous and allogenic tissue engineered leaflets were implanted in lambs to replace the right posterior leaflet of the pulmonary valve with an engineered valve leaflet. Results at 4 weeks of implantation showed that all animals survived the procedure. There was no evidence of stenosis and trivial pulmonary regurgitation in the autografts, with only moderate regurgitation in the allogenic valves. Collagen analysis of the constructs showed that the implanted valve developed an extracellular matrix. Histologic evaluation of the tissue engineered valve demonstrated that the valve had appropriate cellular architecture. The authors concluded that a tissue engineered valve leaflet constructed from its cellular components could function in the pulmonary valve position. Tissue engineering of a heart valve leaflet is feasible. The author also suggested that, based on their experimental results, autograft tissue will probably be superior to allogenic tissue (Shinoka et al., 1995).

The function and behavior of the seeded cells were further studied in a lamb model. The tissue engineered heart valves were constructed in a similar way as described above. The constructed autologous cell/polymer constructs were implanted in six lambs. In two additional control animals, a leaflet of polymer was implanted without prior cell seeding. The animals were killed either after 6 hours or after 1, 6, 7, 9, or 11 weeks, and the implanted valve leaflets were examined histologically, biochemically, and biomechanically. The results showed that the acellular polymer leaflets were completely degraded, with no residual leaflet tissue at 8 weeks. The tissue-engineered valve leaflet persisted in each animal in the experimental group. The constructs showed a progressive increase in collagen content and the presence of elastin fibers in the matrix. It was concluded that the transplanted autologous cells generated a proper matrix on the polymer scaffold in a physiological environment at a period of 8 weeks after implantation (Shinoka et al., 1996).

In another animal experiment, three-leaflet, valved, pulmonary conduits were constructed from autologous ovine vascular cells and biodegradable polymers. Composite scaffolds of polyglycolic acid and polyhydroxyoctanoates were formed in a conduit, and three leaflets (polyhydroxyoctanoates) were sewn into the conduit. These constructs were seeded with autologous medial cells on 4 consecutive days and coated once with autologous endothelial cells. Thirty-one days after cell harvesting, eight seeded and one unseeded control constructs were implanted to replace the pulmonary valve and main pulmonary artery on a cardiopulmonary bypass in sheep. No postoperative anticoagulation was given. Experimental results showed that the seeded constructs demonstrated increasing cellular and extracellular matrix content with no thrombus formation. The unseeded construct developed thrombus formation on all three leaflets after 4 weeks. This experimental study also showed that valved conduits constructed from autologous cells and a biodegradable matrix can function in pulmonary circulation. The progressive cellular and extracellular matrix formation indicates that remodeling of the tissue engineered structure continues for at least 6 months (Stock et al., 2000).

A crucial factor in the tissue engineering of heart valves is the type of scaffolding material. A study was performed to study three different commonly used biodegradable scaffolding materials, polyglycolic acid (PGA), polyhydroxyalkanoate (PHA), and poly-4-hydroxybutyrate (P4HB), used as scaffolds for tissue engineering of heart valves in a pulsatile flow bioreactor. The cell attachment to these polymers was evaluated after seeding each sample with ovine vascular cells and incubating the cell – polymer construct for 8 days. The seeded vascular constructs were then exposed to continuous flow for 1 hour. Experimental results showed that after seeding and incubating the PGA-, PHA-, and P4HB-scaffolds, there were significantly ($p < 0.001$) more cells on PGA compared with PHA and P4HB. There were no significant differences among the materials after flow exposure in the bioreactor, although PHA and P4HB also demonstrated a considerable amount of cell attachment and collagen development. There was a significantly higher cell attachment and collagen content on the PGA samples compared with P4HB and PHA (Sodian et al., 2000).

A collagen matrix also was explored as a scaffold material. In one experimental study, a xenogenic acellularized collagen matrix was obtained from porcine aortic valves. Human endothelial cells were isolated in parallel from human saphenous veins and expanded *in vitro*. Specimens of the surface of the acellular matrix were seeded with endothelial cells and maintained in culture for up to 3 days. The *in vitro* experiment showed promising results which led the authors to believe that this technique may lead to engineering of tissue heart valves repopulated with the patients' own autologous cells (Bader et al., 1998).

A three-dimensional fibrin gel structure was also explored as a scaffold material. In an *in vitro* experiment, human myofibroblasts obtained form human aortic tissue were cultured in the fibrin gel scaffold. Results showed homogenous cell growth and confluent collagen production. No toxic degradation or inflammatory reactions could be detected (Ye et al., 2000).

The mechanical stability of tissue engineered heart valves strongly depends on the collagen content. To increase the collagen content in the *in vitro* culture stage prior to implantation, various culture conditions were applied. It has been found that when cultures of human myofibroblasts on a biodegradable scaffold exist, they strained the scaffold, and the use of ascorbic acid in the culture medium significantly increased the collagen content. This concept is a further step toward the creation of a hemodynamically competent autologous heart valve (Hoerstrup et al., 1999).

5.9 Future Challenges

Tissue engineering has made significant progress in the past decade. Great success has been achieved with some tissues and organs, such as skin, bone, and cartilage, thanks to their relative simplicity. The final goal of tissue engineering will be the complete regeneration of internal organs, such as the heart and liver, which perform more complicated physiological and biochemical functions. To achieve this goal, researchers are facing many challenges in the materials, biology, and engineering fields.

Scaffold materials play a critical role in providing mechanical stability to tissue engineered constructs prior to synthesis of a new extracellular matrix by the cells. Therefore, it is desirable to have materials which match the mechanical properties of the tissue. As an example, scaffolds for bone (trabecular bone) tissue engineering should have high stiffness and high strength. Though considerable efforts have been made to develop high strength, high stiffness yet biodegradable materials, no satisfactory materials are available yet.

Regeneration of some organs needs more sophisticated scaffolds. These scaffolds have to play a more active role in guiding tissue development. New biomaterials have to be designed to encourage cell attachment and proliferation. To achieve this goal, polypeptides can be covalently attached to the surface of currently available biomaterials or integrated into the backbone of polymeric biomaterials. The real challenge would be to create synthetic new biomaterials which mimic the extracellular matrix and induce direct attachment of cells to a scaffold.

The tissue regeneration process involves the production of an extracellular matrix, the use of factors, and the recovery of the functions of the tissue *in vivo*. The whole regeneration process remains largely unexplored, for example, the regulation of matrix production, pattern formation, and morphogenesis (Bonassar and Vacanti, 1998). Thoroughly understanding the process will lead to a higher level in tissue engineering controlling the critical events in the tissue regeneration process and guiding tissue formation more effectively.

To transfer a lab technology to a clinical setting, scale-up remains a large problem, especially in the large-scale expansion of cells. Bioreactors present numerous advantages for growing large amounts of tissue quickly. These advantages are that bioreactors provide uniform mixing and precise control over mass transfer rates and pH and facilitate maintenance of nutrient levels. In addition, some external stimuli, such as electric and mechanical stimuli, can be applied. Due to the complexity of engineered tissues, bioreactors have to be specifically designed to fulfill the special needs of tissue culture.

References

Aebischer, P., R. F. Valentini, P. Dario, C. Domenici, P. M. Galletti. Brain Res. 436:165 (1987)

Aebischer, P., V. Gunard, S. Brace. J. Neurosci. 9:3590 (1989a)

Aebischer, P., A. N. Salessiotis, S. R. Winn. J. Neuroscience Res. 23:282 (1989b)

Aebischer, P., V. Guenard, R. F. Valentini. Brain Res. 531:211 (1990)

Albrektsson, T., G. Zarb, P. Worthington, A. S. Ericsson. Int. J. Oral Maxillof acial Impl ants. 1:11 (1986)

Allcock, H. R., S. Kwon, G. H. Riding, R. J. Fitzpatrick, J. L. Bennett. Biomaterials 9:509 (1988)

Allcock, H. R., S. Kwon, S. R. Pucher. Polym. Prepr., Am. Chem. Soc., Div. Polym. Chem. 31:180 (1990)

Allcock, H. R., J. Y. Chang. Macromolecules 24:993 (1991)

Antonucci, J. M., B. O. Fowler, S. Venz. Dent. Mater. 7:124 (1991)

Attawia, M. A., K. Herbert, K. Uhrich, R. Langer, C. Laurencin. J. Biomed. Mater. Res. 48:322 (1999)

Bader, A., T. Schilling, O. E. Teebken, G. Brandes, T. Herden, G. Steinhoff, A. Haverich. Eur. J. Cardio – Thoracic Surg. 14:279 (1998)

Bakker, D., J. R. de Wijn, C. M. F. Vrouenraets, S. C. Hesseling, J. J. Grote, C. A. van Blitterswijk. In Polymers in Medicine and Surgery. 11(1989)

Battiston, B., P. Tos, T. R. Cushway, S. Geuna. Microsurgery 20:32 (2000)

Behiri, J. C., M. Braden, S. N. Khorasani, D. Wiwattanadate, W. Bonfield. Bioceramics 4:301(1991)

Bell, E., H. P. Ehrlich, D. J. Buttle, T. Nakatsuji. Science 211:421 (1981)

Bell, E., N. Parenteau, R. Gay, C. Nolte, P. Kemp, P. Bilbo, B. Ekstein, E. Johnson. Toxic. *in vitro*. 5:591 (1991)

Black, J. *Biological Performence of Materials*. Marcel Dekker, New York (1981)

Boeree, N. R., J. Dove, J. J. Cooper, J. Knowles, G. W. Hastings. Biomaterials

14:793 (1993)

Bonassar, L. J., C. A. Vacanti. J. Cellular Biochem. Suppl. 30/31:297 (1998)

Bonfield, W., J. C. Behiri, C. Doyle, J. Bowman, J. Abram. In: *Biomaterials and Biomechanics*, eds. G. van der Perre, P. Ducheyne, A. E. Aubram. Elsevier Science, Amsterdam, p. 421 (1984)

Bonfield, W., C. Doyle, K. E. Tanner. In: *Biological and Biomedical Performance of Biomaterials.* eds. P. Christel, A. Meunier, A. J. C. Lee. Elsevier Science, Amsterdam, p. 153 (1986)

Bowen, R. L.. J. Am. Dent. Assoc. 66:71 (1963)

Boyce, S. T., J. Christianson. Surgery 103:421 (1988)

Boyce, S. T., J. Christianson, J. F. Hansbrough, J. Biomed. Mater. Res. 22:939 (1989)

Breitbart, A. S., D. A. Grande, R. Kessler, J. T. Ryaby, R. J. Fitzsimmons, R. T. Grant. Plast. And Reconstr. Surg. 101:567 (1998)

Bruder, S. P., K. H. Kraus, V. M. Goldberg, S. Kadiyala. J. Bone J. Surg., Am. 80:985 (1998)

Bryan, D. J., P. L. Eby, P. D. Costas, K. Wang, B. P. Chakalis, B. R. Seckel. *American College of Surgeons* 1992 *Surgical Forum.* XLIII: 651 (1992)

Burke, J. F., I. V. Yannas, W. C. Quinby, C. C. Bondoc, W. K. Jung. Ann. Surg. 194:413 (1981)

Caplan, A. T. Biomaterials 11:44 (1990)

Castaldini, A., A. Cavallini, A. Moroni, R. Olmi, In: *Biomaterials and Biomechanics* 1983, ed. P. Ducheyne, G. van der Perre, A. E. Aubert. Elsevier Science, Amsterdam, p. 427 (1984)

Castaldini, A., A. Cavallini, In: P. Christel, A. Meunier, A. J. C. Lee, eds. *Biological and Biomedical Performance of Biomaterials.* Elsevier Science, Amsterdam, p. 525 (1986)

Castaldini, A., A. Cavallini, D. Cavalcoli, In: A. Pizzoferrato et al., eds. *Biomaterials and Clinical Applications.* Elsevier Science, Amsterdam, p. 517 (1987)

Cataltepe, O., O. E. Ozan, R. Onur, B. Demirhan, S. Ruacan, A. Erbengi. Acta Neurochir. 121:181 (1993)

Chang, A., I. V. Yannas, S. Perutz, H. Loree, R. R. Sethi, C. Krarup, T. V. Norregaard, N. T. Zervas, T. Silver. In: C. G. Gebelein, R. L. Dune, eds. *Progress in Biomedical Polymers.* Plenum Press, New York, p. 107 (1990)

Chistolini, P., I. Ruspantini, P. Bianco, A. Corsi, R. Cancedda, R. Quarto. J. Mater. Sci. Mater. Med. 10:739 (1999)

Colin, W., R. B. Donoff. J. Dent. Res. 63:987 (1984)

Collier, J. H., J. P. Camp, T. W. Hudson, C. E. Schmidt. J. Biomed. Mater. Res. 50:574 (2000)

Compton, C.. Wounds: A Compendium Clin. Res. Pract. 5:97 (1993)

Cook, S. C., G. C. Baffers, M. Wolfe, K. Sampath, D. C. Rueger, T. S. Whitecloud. J. Bone Jt. Surg. 1994:827 (1994)

Cordeiro, P. G., B. R. Seckel, S. A. Lipton, P. A. D'Amore, J. Wagner, R. Madison. Plast. Reconstr. Surg. 83:1013 (1989)

Coulomb, B., P. Saiga, B. Bertraux, C. Lebreton, M. Haslan, E. Bell, L. Dubertret. In: P. Christel, A. Meunier, and A. J., Lee, eds. *Biological and Biomechanical Performance of Biomaterials*. Elsevier, Amsterdam, 177 (1986a)

Coulomb, B., P. Saiga, E. Bell, F. Breitburd, C. Lebreton, M. Haslan, L. Dubertret. Br. J. Dermatol. 114:91 (1986b)

Couves, L. D.. Synth. Met. 28: C761 (1989)

Crommen, J. H. L., E. H. Schacht, E. H. G. Mense. Biomaterials 13:511 (1992a)

Crommen, J. H. L., E. H. Schacht, E. H. G. Mense. Biomaterials 13:601 (1992b)

Dagalaskis, N., J. Flink, P. Stasikelis, J. F. Burke, I. V. Yannas. J. Biomed. Mater. Res. 14:511 (1980)

Daniels, A. U., M. K. O. Chang, K. P. Andriano. J. Appl. Biomater. 1:57 (1990)

Deb, S., M. Wang, K. E. Tanner, W. Bonfield. J. Mater. Sci. Mater. Med. 7:191 (1996)

de Bruijn, J. D., v. d. Brink, I. S. Mendes, R. Dekker, Y. Bovell, C. A. van Blitterswijk. Adv. Dent. Res. 13:7481 (1999)

de Bruijn, J. D., I. van den Brink, Y. V. Bovell. Bioceramics 9:45 (1996)

de Groot, K. In: D. F. Williams, ed., *Biocompatibility of Implant Materials*. CRC Press, Boca Raton, FL, p. 199 (1981)

de Groot, K., R. G. T. Geesink, C. P. A. T. Klein, P. Serekian. J. Biomed. Mater. Res. 21:1375 (1987)

Dean, D., N. S. Topham, C. Rimnac, A. G. Mikos, D. P. Goldberg, K. Jepsen, R. Redtfeldt, Q. Liu, D. Pennington, R. Ratcheson. Plast. Reconstr. Surg. 104:705 (1999)

den Dunnen, W. F. A., M. F. Meek, D. W. Grijpma, P. H. Robinson, J. M. Schakenraad. J. Biomed. Mater. Res. 51:575 – 585 (2000)

den Dunnen, W. F. A., J. M. Schakenraad, G. J. Zondervan, A. J. Pennings, B. van der Lei, P. H. Robinson. J. Mater. Sci. Mater. Med. 4:521 (1993a)

den Dunnen, W. F. A., B. van der Lei, J. M. Schakenraad, E. H. Blaauw, I. Stokroos, A. J. Pennings, P. H. Robinson. Microsurgery 14: 508 – 515 (1993b)

Devin, J., M. Attawia, C. Laurencin. J. Biomed. Mater. Sci. Polym Ed. 7:661 (1996)

Domb, A., S. Amselem, J. Shah, M. Maniar. Polym. Adv. Technol. 3:279 (1992)

Doyle, C., K. E. Tanner, W. Bonfield. Biomaterials 12:841 (1991)

Ducheyne, P., L. L. Hench, I. I. A. Kagan, M. Martens, A. Bursens, J. C. Mulier., J. Biomed. Mater. Res. 14:225 (1980)

Dunn, R. L.. In: J. O. Hollinger, ed., *Biomedical Applications of Synthetic Biodegradable Polymers*. CRC Press, Boca Raton, p. 20 (1995)

Eming, S. A., M. L. Yarmush, J. R. Morgan. Biotechnol. Bioeng. 52: 15 (1996)

Endo, K. Dental Mater. J. 14:185 (1995)

Ferris, D. M., G. D. Moodie, P. M. Dimond, C. W. D. Gioranni, M. G. Ehrlich,

R. F. Valentini. Biomaterials 20:2323 (1999)
Furnish, E., C. Schmidt. In: *Frontiers in Tissue Engineering*, eds. C. W. Patrick, A. G. Mikos, L. V. McIntire. Elsevier Science, New York, p. 514 (1998)
Gallico, G. G. I., N. E. O'Connor, C. C. Compton, O. Kehinde, H. Green. N. Engl. J. Med. 311:448 (1984)
Garner, B., A. Georgevich, A. J. Hodgson, L. Liu, G. G. Wallace. J. Biomed. Mater. Res. 44:121 (1999)
Gearhart, J. Science 282:1061 (1998)
Green, H., O. Kehinder, J. Thomas. Proc. Natl. Acad. Sci. USA. 76:5665 (1979)
Grijpma, D. W., G. J. Zondervan, A. J. Pennings. Polym Bull. 25:327 (1991)
Guida, G., V. Riccio, S. Gatto, C. Migliaresi, L. Nicodemo, L. Nicolais, C. Palomba. In: P. Ducheyne, G. van der Perre and A. E. Aubert, eds. Biomaterials and Biomechanics 1983. Elsevier Science, Amsterdam, p. 19 (1984)
Hansbrough, J. F., B. S. T, M. L. Cooper, T. J. Foreman. JAMA 262:2125 (1989)
Hansbrough, J. F., M. L. Cooper, R. Cohen. Surgery 111:438 (1992a)
Hansbrough, J. F., C. Dore, W. B. Hansbrough. J. Burn Care Rehabil. 13:519 (1992b)
Harvey, A. R., M. Chen, G. W. Plant, S. E. Dyson. Restor. Neurol. Neurosci. 6:221 (1994)
Heimbach, D., A. Leterma, J. F. Burke, A. Cram, D. Herndon, J. Hunt, M. Jordan, W. McManus, L. Solam, G. Warden. Ann. Surg. 208:313 (1988)
Hench, L. L., R. J. Splinter, W. C. Allen, T. K. Greenlee. J. Biomed. Mater. Res. 2:117 (1971)
Higashi, S., T. Yamamuro, T. Nakamura, Y. Ikada, S. H. Hyon, K. Jamshidi. Biomaterials 7:183 – 187 (1986)
Hing, K. A., W. Bonfield. Patent WO19991005. 99-gb3283 (2000)
Hoerstrup, S. P., G. Zund, Q. Ye, A. Schoeberlein, A. C. Schmid, M. I. Turina, ASAIO J. 45:397 (1999)
Hollinger, J. O., K. Leong, Biomaterials 17:187 (1996)
Hulspas, R., C. Tiarks, J. Reilly, C. C. Hsieh, L. Recht, P. J. Quesenberry. Exp. Neurol. 148:147 (1997)
Ishaug-Riley, S. L., G. M. Crane, M. J. Miller, A. W. Yasko, M. J. Yaszemski, A. G. Mikos. J. Biomed. Mater. Res 36:17 (1997)
Jansen, J. A., J. E. de Ruijter, P. T. M. Jansen, Y. G. C. J. Paquay. Biomaterials 16:819 (1995)
Jones, D. W., A. S. Rizkalla. J. Biomed. Mater. Res. (Appl. Biomater.) 33:89 (1996)
Jones, N. L., D. F. Williams. . Toronto, p. II - 441 (1996)
Kadiyala, S., N. Jaiswal, S. P. Bruder. Tissue Eng. 3:173 (1997)
Klein, C., K. de Groot, W. Chen, Y. Li, X. Zhang. Biomaterials 15:31 (1994)
Knowles, J. C., G. W. Hastings, H. Ohta, S. Niwa, N. Boeree. . Biomaterials 13:491 (1992)
Kohn, J., R. Langer. In: B. D. Ratner, A. S. Hoffman, F. J. Schoen, J. E.

Lemons. eds, Biomaterials Science, An Introduction to Materials and Medicine. Academic Press, San Diego, p. 64 (1996)

Kokubo, T., S. Ito, T. Huang, S. Sakka, T. Kitsugi, T. Yamamuro. J. Biomed. Mater. Res. 24: 331 (1990)

Kuijer, R., S. J. M. Bouwmeester, M. M. W. E. Drees, D. A. M. Surtel, E. A. W. Terwindt-Rouwenhorst. A. J. Van der Linden, V. B. C. A., S. K. Bulstra. J. Mater. Sci. Mater. Med. 9:449 (1998)

Labella, R., M. Braden, S. Deb. Biomaterials 15:1197 (1994)

Lacefield, W. R. In: J. E. L. P. Ducheyne, ed. *Bioceramics: Material Characteristics Versus in Vivo Behavior*. The New York Academy of Science, New York, p. 72 (1988)

Langone, F., S. Lora, F. M. Veronese, P. Caliceti, P. P. Parnigotto, F. Valenti, G. Palma. Biomaterials 16:347 (1995)

Laurencin, C. T., M. A. Attawia, H. M. Elgendy, M. Fan. Mater. Res. Soc. Symp. Proc. 414:157 (1996a)

Laurencin, C. T., S. F. El-Amin, S. E. Ibim, D. A. Willoughby, M. Attawia, H. R. Allcock, A. A. Ambrosio. J. Biomed. Mater. Res. 30:133 (1996b)

Laurencin, C. T., M. D. Borden, A. A. Ambrosio, M. A. Attawia, F. K. Ko, H. R. Allcock, G. M. Morrill. *Book of Abstracts*, 216*th ACS Nat. Meet*. Boston, August, 23-27:OLY (1998a)

Lazarus, H. M., S. E. Haynesworth, S. L. Gerson, N. S. Rosenthal, A. I. Caplan. Bone Marrow Transplant. 16:557 (1995)

LeGeros, R. Z. *Calcium Phosphates in Oral Biology and Medicine*. Karger, Basel, Switzerland (1991)

Levi, A. D. O., V. Guenard, P. Aebischer, R. P. Bunge. J. Neurosci. 14:1309 (1994)

Li, Y., J. R. de Wijn, C. P. A. T. Klein, S. v. d. Meer, K. de Groot. J. Mater. Sci. Mater. Med. 5:252 (1994)

Li, H. C., E. Khor. Macromol. Chem. Phys. 196:1801 (1995)

Li, S. H., Q. Liu, J. R. de Wijn, B. L. Zhou, K. de Groot. Biomaterials 18:389 (1997)

Li, S. H., Z. Zheng, Q. Liu, J. R. de Wijn, K. de Groot. J. Biomedical Materials Res. 40:520 (1998)

Lind, M., S. Overgaard, T. Nguyen, B. Ongpipattanakul, C. Bunger, K. Soballe. Acta Orthoped. Scand. 67:611 (1996a)

Lind, M., S. Overgaard, B. Ongpipattanakul, T. Nguyen, C. Bunger, K. Soballe. J. Bone Jt. Surg. 78:377 (1996b)

Liu, Q., J. R. de Wijn, D. Bakker, C. A. van Blitterswijk. J. Mater. Sci. Mater Med. 7:551 (1996)

Liu, Q., J. R. de Wijn, C. A. van Blitterswijk. Biomaterials 18:1263 (1997)

Liu, Q., J. R. de Wijn, D. Bakker, M. v. Toledo, C. A. Van Blitterswijk. J. Mater. Sci. Mater Med. 9:23 (1998a)

Liu, Q., J. R. de Wijn, C. A. van Blitterswijk. J. Biomed. Mater. Res. 40:490 (1998b)

Liu, Q., J. Weng, J. G. C. Wolke, C. A. van Blitterswijk. Cells Mater. 7:41 (1998c)

Liu, Q., J. R. de Wijn, K. de Groot, C. A. van Blitterswijk. Biomaterials 19: 1067 (1998d)

Liu, Q., J. R. de Wijn C. A. van Blitterswijk. J. Biomed. Mater. Res. 40:358 (1998e)

Liu, Q., J. R. de Wijn, C. A. van Blitterswijk. J. Biomed. Mater. Res. 40:257 (1998f)

Liu, Q., E. L. Hedberg, Z. Liu, R. Meszlenyi, R. Bahulekar, A. G. Mikos. Biomaterials 21:2163 (2000)

Livingston, T., J. Garino, P. Ducheyne. Bioceram. *Proc. Int. Symp. Ceram. Med.* 12:245 (1999)

Lu, L., A. G. Mikos. MRS Bull. Nov.:28 (1996)

Luklinska, Z. B., W. Bonfield. J. Mater. Sci. Mater. Med. 8:379 – 383 (1997)

Lundborg, G., H. A. Hansson. Hand Surg. 5:35 (1980)

Marden, L. J., J. O. Hollinger, A. Chaudhari, T. Turek, R. Schaub, E. Ron. J. Biomed. Mater. Res. 28:1127 (1994)

Meek, M. F., W. F. A. Den Dunnen, J. M. Schakenraad, P. H. Robinson. Microsurgery 17:555 (1996)

Meijer, G. J., J. Heethaar, M. S. Cune, C. De Putter, C. A. Van Blitterswijk. Inter. J. Oral Maxillof a cial Surg. 26:135 (1997)

Meijer, G. J., F. J. M. Starmans, C. Deputter, C. A. Van Blitterswijk. J. Oral Rehabil. 22: 105 (1995)

Merle, M., A. L. Dellon, J. N. Campbell, P. S. Chang. Microsurgery 10:130 – 133 (1989)

Miller, L. L., Q. X. Zhou. Macromolecules 20:1594 (1987)

Molander, H., Y. Olsson, O. Engkvist, S. Bowald, I. Eriksson. Muscle Nerve 5:54 (1982)

Nanci, A., J. D. Wuest, L. Peru, P. Brunet, V. Sharma, S. Zalzal, M. D. McKee. J. Biomed. Mater. Res. 40:324 (1998)

Nazhat, S. N. , R. Smith, S. Deb, M. Wang, K. E. Tanner, W. Bonfield, *Trans. 5th World Biomater. Congr.* Toronto, p. II-83 (1996)

Nimni, M. E. Biomaterials 18:1201 (1997)

Nusgens, B., C. Merril, C. Lapiere, E. Bell. Collagen Relat. Res. 4:351(1984)

Ohgushi, H., A. I. Caplan. J. Biomed. Mater. Res. 48:913 (1999)

Ohgushi, H., Y. Dohi, S. Tamai, S. Tabata. J. Biomed. Mater. Res. 27:1401 (1993)

Ohgushi, H., V. M. Goldberg, A. I. Caplan. Acta Orthoped Scand. 60: 334 (1989)

Oonishi, H., L. L. Hench, J. Wilson, F. Sugihara, E. Tsuji, M. Matsuura, S. Kin, T. Yamamoto, S. Mizokawa. J Biomed Mater Res. 51:37 – 46 (2000)

Pachence, J. M. J. Biomed. Mater. Res. 33:35 (1996)

Peleo, D. A. J. Biomed. Mater. Res. 29:951 (1995)

Peleo, D. A. J. Biomed. Mater. Res. 37:222 (1997)

Peter, S. J., M. J. Miller, M. J. Yaszemski, A. G. Mikos. In: A. J. Domb, J. Kost, D. M. Wiseman, eds. *Handbook of Biodegradable Polymers.* Hardwood Academic, Amsterdam, p. 87 (1997)

Peter, S. J., M. J. Miller, A. W. Yasko, M. J. Yaszemski, A. G. Mikos. J.

Biomed. Mater. Res. 43:422 (1998a)
Peter, S. J., S. T. Miller, Z. G., A. Yasko, A. G. Mikos. J. Biomed. Mater. Res. 41:1 (1998b)
Phillips, T. J., O. Kehinde, H. Green, B. A. Gilchrest. J. American Academy of Dermatol. 21:191 (1989)
Plueddemann, E. P. *Silane Coupling Agents*, 2nd Ed., Plenum Press, New York, p. 118 (1991)
Prezyna, L. A., Y. J. Qiu, J. R. Reynolds, G. E. Wnek. Macromolecules 24: 5283 (1991)
Puleo, D. A., W. W. Huh, S. S. Duggirala, P. P. DeLuca. J. Biomed. Mater. Res. 44:104 (1998)
Radder, A. M., J. E. Davies, R. N. S. Sohdi, S. A. T. van de Meer, J. G. C. Wolke, C. A. van Blitterswijk. Bioceramics 6:345 (1993)
Radder, A. M., H. Leenders, C. A. van Blitterswijk. J. Biomed. Mater. Res. 28: 141 (1994)
Reddi, A. H.. Nature Biotechnol. 16:247 (1998a)
Restrepo, Y., M. Merle, J. Michon, B. Folliguet, D. Petry. Microsurgery 4:105 (1983)
Rheinwald, J. G., H. Green. Cells 6:331 (1975)
Ripamonti, U.. J. Bone Jt. Surg. 73A:692 (1991)
Ripamonti, U., S. Ma, A. H. Reddi. Matrix 12:202 (1992)
Robinson, P. H., B. van der Lei, H. J. Hoppen, J. W. Leenslag, A. J. Pennings, P. Nieuwenhuis. Microsurgery 12:412 (1991)
Rodriguez, A. M., C. A. Vacanti. In: C. W. Patrick, A. G. Mikos, L. V. McIntire, eds, *Frontiers in Tissue Engineering*. Elsevier Science, New York, p. 400 (1998)
Rozema, F. R., W. C. de Bruijn, R. R. M. Bos, G. Boering, A. J. Nijenhuis, A. J. Pennings In: P. J. Doherty, R. L. Williams, D. F. Williams, eds. *Biomaterial-tissue interfaces: Advances in Biomaterials.* 10: 349 Elsevier Science, Amsterdam (1992)
Sadik, O. A., G. G. Wallace. Anal. Chim. Acta 279:209 (1993)
Scheridan, R. L., R. G. Tompkins. J. Trauma Injury Infection Criti. Care 38:48 (1995)
Schmidt, C. E., V. R. Shastri, J. P. Vacanti, R. Langer. Proc. Natl. Acad. Sci. USA 94:8948 (1997)
Schoen, F. J., R. J. Levy. J. Biomed. Mater. Res. 47:439 – 465 (1999)
Shastri, V. R., C. E. Schmidt, T. H. Kim, J. P. Vacanti, R. Langer. Mater. Res. Soc. Symp. Proc.. 414:113 (1996)
Shin, M. C., H. S. Kim. Biosensors Bioelectronics 11:161 – 169 (1996)
Shinoka, T., C. K. Breuer, R. E. Tanel, G. Zund, T. Miura, P. X. Ma, R. Langer, J. P. Vacanti, J. E. J. Mayer. Ann. Thorac. Surg. 60:S513 (1995)
Shinoka, T., P. X. Ma, D. Shum-Tim, C. K. Breuer, R. A. Cusick, G. Zund, R. Langer, J. P. Vacanti, J. E. J. Mayer. Circulation 94: II-164 (1996)
Shriver, D. F., J. S. Tonge, A. Barriola, P. M. Blonsky, H. R. Allcock, S. Kwon, P. Austin. Polym. Prepr. Am. Chem. Soc., Div. Polym. Chem. 28: 438 (1987)

Sinibaldik, K., H. Rosen, S. K. Lin, M. De Angelis. Clin. Orthopred Relat Res.. 118:257 (1976)

Skalak, R., C. F. Fox. *Frontiers in Tissue Engineering*, eds. C. W. Patrick, A. G. Mikos, L. V. McIntire. Elsevier Science, New York, p. xv (1998)

Small, I. A., D. Misiek. J. Oral Maxillofacial Surg.. 44:60 (1986)

Sodian, R., S. P. Hoerstrup, J. S. Sperling, D. P. Martin, S. Daebritz, J. E. J. Mayer, J. P. Vacanti. ASAIO J. 46:107 (2000)

Steflik, D. E., A. L. Sisk, G. R. Parr, L. K. Gardner, P. J. Hanes, F. T. Lake, D. J. Berkery, P. Brewer. J. Biomed. Mater. Res. 27:791 (1993)

Stock, U. A., M. Nagashima, P. N. Khalil, G. D. Nollert, T. Herden, J. S. Sperling, A. Moran, J. Lien, D. P. Martin, F. J. Schoen, J. P. Vacanti, J. E. J. Mayer. J. Thorac. Cardiovasc. Surg.. 119:732 (2000)

Tenhuisen, K. S., P. W. Brown, C. S. Reed, H. R. Allcock. J. Mater. Sci. Mater. Med. 7:673 (1996)

Terjesen, T., K. Apalset. J. Orthoped. Res. 6:292 (1988)

Teumer, J., J. Hardin-Young, N. L. Parenteau. In: *Frontiers in Tissue Engineering*, eds. C. W. J. Patrick, A. G. Mikos, L. V. Mcintire. Elsevier Science, New York, p. 664 (1998)

Thomson, J. A., J. Itskovitz-Eldor, S. Shapiro, M. A. Waknitz, J. J. Swiergiel, V. S. Marshall, J. M. Jones. Science 282:1145 (1998)

Thomson, R. C., A. G. Mikos, E. Beahm, J. C. Lemon, W. C. Satterfield, T. B. Aufdemorte, M. J. Miller. Biomaterials 20:2007 (1999)

Thomson, R. C., M. J. Yaszemski, T. P. Harrigan, J. M. Powers, A. G. Mikos. Mater. Res. Soc. Symp. Proc. 394:25 (1995)

Toriumi, D. M., H. S. Kotler, D. P. Luxunberg, M. E. Holtrop, E. A. Wang. Arch. Otolaryngol. Head Neck Surg. 117:1101 (1991)

Vacanti, C. A., W. Kim, J. Upton, D. Mooney, J. P. Vacanti. Tissue Eng. 1: 301 (1995)

Valentini, R. F., T. G. Vargo, Jr. Gardella, J. A., P. Aebischer. Biomaterials. 13:183 (1992)

van Blitterswijk, C. A., D. Bakker, H. Lenders, J. v. d. Brink, S. C. Hesseling, Y. P. Bovell, A. M. Radder, R. J. Sakker, M. L. Gallard, P. H. Heinze, G. J. Beumer. In: T. K. P. Ducheyne, C. A. van Blitterswijk, eds. *Bone-Bonding Biomaterials*. Reed Healthcare Communications, Leiderdorp, The Netherlands, p. 153 (1992)

van Blitterswijk, C. A., J. V. D. Brink, H. Leenders, D. Bakker. Cells Mater. 3: 23 (1993)

van der Meer, S. A. T., J. R. de Wijn, J. G. C. Wolke. J. Mater. Sci. Mater. Med. 7:359 (1996)

van Dorp, A. G. M., M. C. H. Verhoeven, H. K. Koerten, T. H. van der Meij, C. A. van Blitterswijk, M. Ponec. Wound Rep. Reg. 6:556 (1998)

van Dorp, A. G. M., M. C. H. Verhoven, H. K. Koerten, C. A. van Blitterswijk, M. Ponec. J. Biomed. Mater. Res. 47:292 (1999)

van Raemdonck, W., P. Ducheyne, P. DeMeester. J. Am. Ceram. Soc. 6:381 (1984)

van Sliedregt, A., K. de Groot, C. A. vna Blitterswijk. J. Mater. Sci: Mater. in

Med. . 4:213 (1993)
van Sliedregt, A., A. M. Radder, K. de Groot, C. A. van Blitterswijk. J. Mater. Sci. Mater. Med. 3:365 (1992)
Venhoven, B. A. M., A. J. de Gee, A. Werner, C. L. Davidson. Biomaterials 15: 1152 (1994)
Verheyen, C. C. P. M., J. R. de Wijn, C. A. van Blitterswijk, P. M. Rozing, K. de Groot. In: P. Ducheyne, T. Kokubo, C. A. van Blitterswijk, eds. Bone-bonding Biomaterials. Reed Healthcare Communications, Leiderdorp, The Netherlands, p. 153 (1992a)
Verheyen, C. C. P. M., J. R. de Wijin, C. A. van Blitterwijk, K. de Groot. J. Biomed. Mater. Res. 26: 1277 – 1296 (1992b)
Verheyen, C. C. P. M., J. R. de Wijn, C. A. van Blitterswijk, K. de Groot, P. M. Rozing. J. Biomed. Mater. Res. 27:433 (1993)
Walton, R. L., R. E. Brown, E. Matory, G. L. Borah, J. L. Dolph. Plast Reconstr. Surg. 84:944 (1989)
Wang, K. K., P. D. Costas, D. J. Bryan, D. S. Jones, B. R. Seckel. Microsurgery 14:608 (1993a)
Wang, K. K., P. D. Costas, D. S. Jones, R. A. Miller, B. R. Seckel. J. Reconstr. Microsurgery 9:39 (1993b)
Wang, M., D. Porter, W. Bonfield. Br. Ceram. Trans. 93:91 (1994)
Weadock, K. W., R. M. Olson, F. H. Silver. Biomater. Med. Devices Artif. Orgrans 11:293 (1984)
Wen, H. B., J. R. de Wijn, Q. Liu, K. de Groot. J. Mater. Sci. Mater. Med. 8:765 (1997)
Wen, H. B., Q. Liu, J. R. de Wijn, K. de Groot, F. Z. Cui. J. Mater. Sci.: Mater. Med. 9:121 (1998)
Widmer, M. S., A. G. Mikos. In: C. W. Patric Jr, A. G. Mikos, L. V. Mcintire, eds. *Frontiers in Tissue Engineering*. Elsevier Science, New York, p. 107 (1998a)
Widmer, M. S., P. K. Gupta, L. Lu, R. K. Meszlenyi, G. R. D. Evans, K. Brandt, T. Savel, A. Gurlek, C. W. Patrick, Jr., A. G. Mikos. Biomaterials 19:1945 (1998b)
Williams, L. R., N. Danielsen, H. Muller, S. Varon. J. Comp. Neurol. 264:284 (1987)
Xiao, Y. L., J. Riesle, C. A. Van Blitterswijk. J. Mater. Sci. Mater. Med. 10: 773 (1999)
Yannas, I. V., J. F. Burke. J. Biomed. Mater. Res. 14:65 (1980a)
Yannas, I. V., J. F. Burke, P. L. Gordon, C. Huang, R. H. Rubenstein. J. Biomed. Mater. Res. 14:107 (1980b)
Yannas, I. V., J. F. Burke, D. P. Orgill, E. M. Skrabut. Science 215: 174 (1982)
Yannas, I. V., J. F. Burke, D. P. Orgill, E. M. Skrabut, G. F. Murphy. Proc. Natl. Acad. Sci. USA 86:933 (1989)
Yannas, I. V.. Angew. Chem. Int. Ed. Engl. 29:20 (1990)
Yannas, I. V.. In: B. D. Ratner, A. S. Hoffman, F. J. Schoen, J. E. Lemons, eds. *Biomaterials Science, An Introduction to Materials and Medicine*.

Academic Press, San Diego, p. 84 (1996)
Yasko, A. W., J. M. Lane, F. E. J., V Rosen, J. M. Wozney, E. A. Wang. J. Bone J. Surg. 74-A:659 (1992)
Yaszemski, M. J., R. G. Payne, W. C. Hayes, R. Langer, A. G. Mikos. Biomaterials 17:175 (1996a)
Yaszemski, M. J., R. G. Payne, W. C. Hayes, R. Langer, A. G. Mikos. Biomaterials 17:2127 (1996b)
Yaszemski, M. J., R. G. Payne, W. C. Hayes, R. S. Langer, A. G. Mikos. Tissue Eng. 1:41 (1995)
Ye, Q., G. Zund, P. Benedikt, S. Jockenhoevel, S. P. Hoerstrup, S. Sakyama, J. A. Hubbell, M. Turina. Eur. J. Cardiothorac. Surg. 17:587 (2000)
Zacchi, V., C. Soranzo, R. Cortivo, M. Radice, P. Brun, G. Abatangelo. J. Biomed. Mater. Res. 40:187 (1998)
Zhang, R., P. Ma. J. Biomed. Mater. Res. 44:446 (1999)

Index

Index

Printed in the United Kingdom
by Lightning Source UK Ltd.
116102UKS00007B/62